METHODS IN MOLECULAR BIOLOGY™

RNA Isolation and Characterization Protocols

Edited by

Ralph Rapley

University of Hertfordshire, Hatfield, UK

and

David L. Manning

Tenovus Institute for Cancer Research, Cardiff, UK

Humana Press ✳ Totowa, New Jersey

572.88 R

CARDIFF UNIVERSITY
PRIFYSGOL CAERDYDD
10 MAY 1999

© 1998 Humana Press Inc.
999 Riverview Drive, Suite 208
Totowa, New Jersey 07512

This publication is printed on acid-free paper. ∞
ANSI Z39.48-1984 (American Standards Institute) Permanence of Paper for Printed Library Materials.

Cover illustration: Fig. 2B from "Analysis of RNA by Northern Blotting Using Riboprobes," by Rai Ajit K. Srivastava.

Cover design by Jill Nogrady.

For additional copies, pricing for bulk purchases, and/or information about other Humana titles, contact Humana at the above address or at any of the following numbers: Tel.: 973-256-1699; Fax: 973-256-8341; E-mail: humana@humanapr.com; Website: http://humanapress.com

Photocopy Authorization Policy:
Authorization to photocopy items for internal or personal use, or the internal or personal use of specific clients, is granted by Humana Press Inc., provided that the base fee of US $8.00 per copy, plus US $00.25 per page, is paid directly to the Copyright Clearance Center at 222 Rosewood Drive, Danvers, MA 01923. For those organizations that have been granted a photocopy license from the CCC, a separate system of payment has been arranged and is acceptable to Humana Press Inc. The fee code for users of the Transactional Reporting Service is: [0-89603-393-7/98 (combbound) 0-89603-494-1/98 (hardcover) $8.00 + $00.25].

Printed in the United States of America. 10 9 8 7 6 5 4 3 2 1

Library of Congress Cataloging in Publication Data

RNA isolation and characterization protocols/edited by Ralph Rapley, David L. Manning.
        p.   cm. – (Methods in molecular biology™; 86)
    Includes index.
    ISBN 0-89603-494-1 (hard: alk. paper). – ISBN 0-89603-393-7 (comb: alk. paper)
    1. RNA–Purification–Laboratory manuals. 2. RNA–Analysis–Laboratory manuals.
    I. Rapley, Ralph.   II. Manning, David L.   III. Series: Methods in molecular biology (Clifton, NJ); 86.
    QP623.R575   1998
    572.8'8–dc21

Preface

Ribonucleic acids are central to cellular and molecular processes and perform vital functions in both structural and functional roles. RNA molecules form the bridge between the stable genetic information contained within DNA and enzymes and proteins that carry out much of the metabolism within the cell. Many of the sites of protein synthesis, the ribosomes within the cell, are composed of these ribonucleic acids as are the tRNA molecules that deliver the amino acid building blocks to the ribosomes. Of all the RNA species, the nucleic acid intermediate, messenger RNA, is a desirable source of material to biologists, since this reflects much of, what ultimately, is translated into enzymes and proteins. In order to determine the qualitative and quantitative changes in mRNA expression, a vast number of molecular biological techniques have been developed.

Key molecular methods that provide the means to initially isolate and analyze RNA molecules are the focus of this volume. In putting together this collection of protocols, we have tried to provide techniques that are most applicable and widely used. In particular, there are a number of isolation techniques included that have been developed, modified, or adapted to enable extraction from a variety of cell types, organisms, or subcellular organelles. Successful isolation of intact RNA is an essential starting point for any subsequent analysis. This is why we have aimed to make this section comprehensive.

The analysis of RNA is the focus of the following chapters. It includes traditional methods of blotting and hybridization, through to those techniques such as differential display, which can measure changes in the expression of specific genes. Readers of this volume will see that many of the methods have been developed through the application of the polymerase chain reaction, which continues to be a profoundly influential technique used today. These and later chapters deal with transcription and translation in vitro and in situ visualization of RNA molecules.

All of the chapters are presented in the familiar style of the Methods in Molecular Biology™ series, with a short description of the basic theory of the technique and an outline of the procedure, followed by a Materials section listing all the reagents necessary for the protocol. The Methods section provides a full and comprehensive account of the protocol in a step-by-step series of actions. In addition, references to the notes provide valuable and useful pieces of information not found in traditional scientific papers, but which in many cases may mean the difference between success and failure of a particular method. All contributors carry out their own research or lead groups that apply the techniques as a matter of routine and therefore are best suited to providing such methods. In putting together *RNA Isolation and Characterization Protocols,* we would like to thank all the present authors that have taken the trouble and valuable time to prepare the individual chapters, Professor John Walker, the series editor for his helpful advice and guidance, and the staff at The Humana Press.

Ralph Rapley
David L. Manning

Contents

Contributors

DAVID B. BATT • *Department of Microbiology, University of Connecticut Health Center, Farmington, CT*

DOMINIQUE BELIN • *Department of Pathology, University of Geneva Medical School, Geneva, Switzerland*

SERENA BONIN • *International Centre of Genetic Engineering and Biotechnology-Padriciano (Trieste), Department of Pathology, University of Trieste*

THOMAS C. G. BOSCH • *Institut der Universitat, Munchen, Germany*

CHARLES D. BOYD • *University of Medicine and Dentistry of New Brunswick, New Brunswick, NJ*

SIAN BRYANT • *Tenovus Cancer Research Centre, University of Wales College of Medicine, Heath Park, UK*

GORDON G. CARMICHAEL • *Department of Microbiology, University of Connecticut Health Center, Farmington, CT*

SHU-HUA CHENG • *Department of Biochemistry, University of Nevada, Reno, NV*

ALAN COLMAN • *School of Biochemistry, Birmingham University, Birmingham, UK*

CHRISTOPHER E. DAHLE • *Department of Medicine, University of Iowa, Iowa City, IA*

ERIC DE KANT • *Academeic Hospital Utrecht, Department of Internal Medicine, Ultrecht, The Netherlands*

ANNE GROBLER-RABIE • *University of Medicine and Dentistry of New Brunswick, New Brunswick, NJ*

JOHN HEPTINSTALL • *Biosciences Group, School of Natural and Environmental Sciences, Coventry University, Coventry, UK*

RACHEL HODGE • *Department of Botany, University of Leicester, Leicester, UK*

MARK LEONARD • *Developmental Biology Research Centre, Kings College, University of London, UK*

ZHONG LIU • *Department of Microbiology, University of Connecticut Health Center, Farmington, CT*

JAN U. LOHMANN • *Institut der Universitat, Munchen, Germany*

RAINER LOW • *Botanisches Institute, Ruprecht-Karls-Universitat, Heidelberg, Germany*

DONALD E. MACFARLANE • *Department of Internal Medicine, University of Iowa, Iowa City, IO*

FRANCOIS MALLET • *Ecole Normale Superieure de Lyon, Lyon, France*

BERNARD MANDRAND • *Ecole Normale Superieure de Lyon, Lyon, France*

DAVID L. MANNING • *Tenovus Cancer Research Centre, University of Wales College of Medicine, Heath Park, UK*

ROBERTO MANTOVANI • *DOI/LGME Faculte de Medicine, Strasbourg, Cedex, France*

LESLEY ANN MARTIN • *ICRF Oncology, Royal Postgraduate Medical School, Surrey, UK*

GLEN MATTHEWS • *Department of Surgery, Queen Elizabeth Hospital, Birmingham, UK*

PETER MERTENS • *Institute for Animal Health, Pirbright, Surrey, UK*

TAPAS MUKHOPADHYAY • *Department of Thoracic and Cardiovascular Surgery, The University of Texas M. D. Anderson Cancer Center, Houston, TX*

LOUISE OLLIVER • *University of Medicine and Dentistry of New Brunswick, New Brunswick, NJ*

GUY ORIOL • *Ecole Normale Superieure de Lyon, Lyon, France*

ROGER PATIENT • *Developmental Biology Research Centre, Kings College, University of London, UK*

ROSELLA PERIN • *International Centre of Genetic Engineering and Biotechnology-Padriciano (Trieste), Department of Pathology, University of Trieste*

ULRICH PFEFFER • *Laboratory of Molecular Biology, National Institute of Genoa, Genoa, Italy*

RALPH RAPLEY • *University of Hertfordshire, Hatfield, UK*

RALF RESKI • *Institute for General Botany, University of Hamburg, Hamburg, Germany*

JACK A. ROTH • *Department of Thoracic and Cardiovascular Surgery, The University of Texas M. D. Anderson Cancer Center, Houston, TX*

ELAINE T. SCHENBORN • *Promega Corporation, Madison, WI*

GUSTOV SCHONFELD • *Department of Internal Medicine, Washington University School of Medicine, St. Louis, MO*

JEFF SEEMAN • *Department of Biochemistry, University of Nevada, Reno, NV*

RAI AJIT K. SRIVASTAVA • *Department of Internal Medicine, Washington University School of Medicine, St. Louis, MO*

GIORGIO STANTA • *International Centre of Genetic Engineering and Biotechnology-Padriciano (Trieste), Department of Pathology, University of Trieste*

ULRIKE STEIN • *Department of Molecular and Tumor Therapy, Max-Delbruck-Centre for Molecular Medicine, Berlin, Germany*

ANU SUOMALAINEN • *National Public Health Institute, Department of Human Molecular Genetics, Helsinki, Finland*

ANN-CHRISTINE SYVANEN • *National Public Health Institute, Department of Human Molecular Genetics, Helsinki, Finland*

GIORGIO TERENGHI • *Blond McIndoe Centre, Queen Victoria Hospital, Sussex, UK*

BIMAL D. M. THEOPHILUS • *Department of Hematology, Birmingham Children's Hospital, Birmingham, UK*

WOLFGANG UCKERT • *Department of Molecular and Tumor Therapy, Max-Delbruck-Centre for Molecular Medicine, Berlin, Germany*

STEPHANE VIVILLE • *DOI/LGME Faculte de Medicine, Strasbourg, Cedex, France*

MAGGIE WALMSLEY • *Developmental Biology Research Centre, Kings College, University of London, UK*

WOLFGANG WALTHER • *Department of Oncology/Surgery, Max-Delbruck-Centre for Molecular Medicine, Berlin, Germany*

FEI YE • *Department of Biology, Massachusetts Institute of Technology, Cambridge, MA*

1

Introduction to Isolating RNA

Donald E. Macfarlane and Christopher E. Dahle

1. Introduction

1.1. Structure

Ribonucleic acid (RNA) is an unbranched polymer of purine (adenine, guanine) and pyrimidine (cytosine, uracil) nucleotides joined by phosphodiester bonds.

The RNA polymer is bulkier than that of DNA (which lacks the 2'OH on the ribose). RNA is usually single-stranded, and it tends to form tertiary structures of high complexity, including hairpin loops, internal loops, bulges, and pseudo-knots (in which self complementary sequences align to form short, antiparallel double helical strands), and triple stranded structures *(1–3)*. This complexity gives rise to functional molecules of much greater diversity than DNA, supporting the provocative concept that RNA evolved in the prebiotic era *(4)*. Intact RNA is difficult to isolate because, as a long polymer, it is prone to mechanical or chemical degradation, and because of the existence of RNase, which is widely distributed on laboratory surfaces and difficult to destroy.

1.2. Function

It is increasingly recognized that RNA molecules act as enzymes as well as serving structural and informational functions. RNA within cells is generally associated with proteins and metal ions in ribonucleoprotein (RNP) complexes *(5)*, of which ribosomes (which synthesize proteins), are the best-known example. RNA is synthesized in precursor form by transcription from DNA.

Transfer RNAs deliver amino acids to ribosomes, and consist of characteristic clover leaf structures about 80 nucleotides long. Many of the base resi-

From: *Methods in Molecular Biology, Vol. 86: RNA Isolation and Characterization Protocols*
Edited by: R. Rapley and D. L. Manning © Humana Press Inc., Totowa, NJ

dues are methylated or otherwise modified. Messenger RNA is capped at its 5'-end with methylated bases, and is usually polyadenylated at its 3'-end. Most eukaryotic mRNAs are spliced from their precursor transcripts to excise introns and juxtapose sequences from exons. mRNAs constitute about $\frac{1}{100}$ of the total RNA of the cell, and yet they carry all genetic information from DNA to the ribosome to generate appropriate sequences of amino acids in the synthesis of proteins. The half-life of mRNAs varies from a few minutes (in the case of eukaryotic regulatory proteins) to years (in seeds and spores). Ribosomal RNA constitutes the bulk of cellular RNA, contributing three molecular species and about half the mass to the organelle which is assembled with more than fifty proteins. Heterogeneous nuclear RNA includes RNA undergoing processing by spliceosomes, in which RNA is itself catalytic.

2. Evolution of Methods for Isolating RNA

2.1. Purposes

The progress of scientific inquiry engenders a coordinated evolution of practical methodology and factual knowledge. As our understanding of RNA has progressed, the amounts and purity of RNA required for experiments has changed. Studies examining the infectivity of viral RNA, or the ability of an RNA to support translation in vitro, demand full-length RNA molecules, but purity (in the chemical sense) is less important than the elimination of interfering molecules. Analysis by ultracentrifugation or by gel electrophoresis and hybridization (Northern blots, typically requiring about 10 μg RNA) requires RNA with a high degree of preservation of polymer length, but these techniques are tolerate of gross impurity and occasional chemical modification of bases (*see* Chapters 11 and 13). Modern methods for measuring the quantity of an known RNA species present in a mixture (such as the RNase protection assay and branched chain analysis and needing up to 50 μg) are tolerant of both impurities and occasional strand breaks, but they require reproducible (and preferably quantitative) recovery of RNA. RNA isolated to prepare cDNA libraries require the highest degree of structural and sequence integrity. RNA intended for amplification by RT-PCR (typically less than 1 μg) must be free of inhibitors of reverse transcriptase or the DNA polymerase, but useful results can often be obtained with impure, degraded samples.

Special methods may be needed to prepare RNA from plants and single-cell organisms to rupture cell walls and to eliminate contaminating (non-nucleic acid) polymers. Methods to prepare RNA in a clinical area cannot employ noxious reagents, and should tolerate prolonged standing at room temperature. For some purposes it may be necessary to isolate RNA free of DNA. Almost all methods demand that the RNA product is free from RNase.

2.2. Early Methods

The earliest methods for isolating RNA were applied to viruses, such as the tobacco mosaic virus, in which the RNA is encapsulated in a protein coat. Brief heating of a suspension of purified viral particles resulted in a coagulum of denatured protein, and a solution of RNA, which was concentrated by dialysis and dried, yielding RNA with molecular weight ranging up to 200,000, a result which challenged the view that RNA was a prosthetic group for a (proteinaceous) enzyme *(6)*. Work with eukaryote RNA was initially directed to subcellular fractions enriched for organelles consisting of ribonucleoproteins, such as ribosomes. During the subcellular fractionation, the RNA was retained in the ribonucleoprotein complex by maintaining a near-physiologic pH, ionic strength and divalent ion concentration, and the detergents used were either non-ionic or a low concentration of an anionic surfactant. These experimental conditions coincidently limited the disruption of the nucleus and the release of DNA. Subsequent differential centrifugation generally resulted in preparations of organelles containing RNA in high concentration. RNA in these purified organelles is relatively protected from RNase.

Following the preparation of the organelles, a variety of methods were used to dissociate the RNA from the protein, and these methods generally included an inhibitor of RNase. Anionic surfactants were particularly useful for this purpose, including sodium dodecyl sulfate (SDS), lithium dodecyl sulfate (which, having a lower Krafft temperature, can be used in the cold *(7)*, and sodium lauroylsarcosine (an amide derivative of SDS used because it is more compatible with cesium chloride centrifugation).

After the dissociation of RNA from the protein, the two can be separated by ultracentrifugation through a cesium chloride gradient ($5.7M$), a very useful technique which exploits the high buoyant density of RNA in cesium chloride (1.9), compared with that of DNA (1.7), and polysaccharides and glycogen (~ 1.67) *(8)*. Pure RNA is sedimented to the bottom of the tube.

Certain organic solvents can dissociate RNA from protein, and by exploiting the difference in hydrophobicity between RNA and protein, can separate them by generating two phases. The most commonly used reagent for this purpose is phenol. After phase separation, the RNA remains in the aqueous phase, whereas proteins (and DNA when conditions are appropriately adjusted) partition either into the phenol layer or collect at the interface.

Methods using phenol can be used to recover RNA directly from whole cells, without prior purification of nucleoprotein particles. In the methods described by Kirby *(9)*, tissues were homogenized in phenol with *m*-cresol (added to reduce the melting temperature of the phenol and to improve its deproteinizing effect), 8-hydroxyquinoline (to chelate metal ions, reduce the rate of oxidation of phenol, and to assist in the inhibition of RNase), and nathphalene 1,5-disulfonate or sodium 4-aminosalicylate (as surfactants). The aqueous phase was collected, and repeatedly re-extracted with a phenol mixture, followed by precipitation of the RNA with ethanol plus either sodium acetate or sodium chloride/sodium benzoate/m-cresol. The yield of rapidly labeling RNA can be increased when the extraction with phenol is carried out at elevated temperature. The addition of chloroform to the phenol often increases the yield of RNA (10).

Methods employing phenol are widely used, but this obnoxious reagent causes much mischief in the hands of the unwary. Phenol (remarkably) does not completely inhibit RNase, and it may actually disrupt the actions of other inhibitors of RNase. It is also prone to oxidize to reactive species which may degrade RNA, requiring that it be purified before use.

2.3. Recent Methods

Chaotropic agents disrupt the forces responsible for cell structure. The use of guanidine hydrochloride (guanidinium chloride) led to the first preparation of eukaryotic RNA in highly polymerized form (11). When used at $4M$, it inhibits RNase and dissociates nucleoproteins. The liberated RNA can be recovered by precipitation with ethanol, or by centrifugation (12).

Guanidine thiocyanate (guanidinium isothiocyanate) is a more powerful chaotrope, and is capable of dissolving most cell constituents, releasing RNA and inhibiting RNase. In the widely used method of Chirgwin (13), which can be applied directly to cells, tissues are homogenized in $4M$ guanidine thiocyanate, $0.1M$ β-mercaptoethanol (a sulphydryl reductant), and 0.5% sodium lauroyl sarcosine. The resulting homogenate is layered on a cesium chloride gradient, and the RNA is ultracentrifuged overnight into a pellet. As an alternative to ultracentrifugation, the RNA can be precipitated with ethanol from the lysis solution, and repeatedly reprecipitated after redissolving in guanidine hydrochloride. The ultracentrifugation method is probably the most reliable method of obtaining high-quality RNA suitable for any purpose, but it is time consuming, and the number of samples that can be processed is limited by the availability of an ultracentrifuge.

The most commonly applied method for isolating RNA in the experimental laboratory uses a proprietary mixture of stabilized phenol and guanidine thiocyanate, into which the sample is homogenized. A chloroform reagent is then added to effect a separation of phases, during which RNA (but not DNA or protein) remains in the aqueous phase, from which it is precipitated (14).

We recently introduced a novel method for isolating RNA from whole cells which takes advantage of the properties of cationic surfactants. These interesting reagents have long been known to precipitate RNA and DNA from aqueous solution. We found that the completeness of this precipitation depended on the nature of the counter ion. We also found that appropriately selected cationic surfactants were capable of lysing cells efficiently, resulting in immediate precipitation of nucleic acids, presumably by the formation of reverse micelles *(15)*. In this state, RNA is protected from RNase. The currently recommended procedure is to homogenize the cells in a solution of $0.1M$ tetradecyltrimethylammonium oxalate, followed by gentle centrifugation. After the supernatant is discarded, the pellet is extracting with $2M$ lithium chloride. RNA, being insoluble in this salt solution, remains in the pellet; but DNA, the surfactant, and some polysaccharides are solubilized. The RNA is then simply dissolved from the pellet with a buffer *(16)*. This simple method avoids obnoxious reagents. Once the sample is mixed with the cationic surfactant, it can be mailed at room temperature to a reference laboratory. These two features are desirable for those planning to explore clinical applications of RNA-based diagnosis in a cost-sensitive age.

3. The Future

The improvements in techniques for isolating RNA that have occurred over the past two decades have materially advanced the progress of molecular biology. RNA isolation is becoming sufficiently reliable to envisage a huge growth in RNA-based diagnostic techniques. Once difficulties in this isolation of RNA have been overcome, RNA will be the most informative class of molecules in the clinical specimen. Informative RNA molecules are usually present in far greater number than the corresponding DNA. Like DNA, RNA reveals the genetic origin of the cell (or virus) containing it, and the analysis of RNA reveals additional information about the activity of the cell at the time that the specimen was collected.

Quite simple methods could be used to detect: invasion by pathogens, tumors with either gene rearrangements or expressing characteristic proteins, inherited disorders caused by altered expression of proteins, and diseases involving the synthesis of proteins characteristic of inflammatory responses. RNA-based methods are currently used to monitor the response to therapy of HIV and hepatitis C infections, and we can anticipate that a similar approach can be applied to a wide variety of disorders. In theory, even differential blood counts and blood typing can be performed using RNA-based methods.

As many readers will recall, isolating RNA used to be a frustrating and tedious task that was a prerequisite to experiments in molecular biology. The chapters in this volume eloquently attest to the advances in RNA isolation and manipulation that have been made over the years. Working with RNA is no longer the ogre it used to be!

References

1. Wyatt, J. R. and Tinoco, I. J. (1993) RNA structure and RNA function, in *The RNA World* (Gesteland, R. F. and Atkins, J. F., eds.), Cold Spring Harbor Laboratory Press, Cold Spring Harbor, NY, pp. 465–496.

2. Choi, Y. C. and Ro-Choi, T. -S. (1980) Basic characteristics of different classes of cellular RNAs: a directory, in *Gene Expression: The Production of RNAs* (Goldstein, L. and Prescott, D. M., eds.), Academic, NY, pp. 609–667.

3. Farrell, R. (1993), in *RNA Methodologies: A Laboratory Guide for Isolation and Characterization.* Academic, San Diego, CA.

4. Gesteland, R. F. and Atkins, J. F. (1993) *The RNA World.* Cold Spring Harbor Laboratory Press, Cold Spring Harbor, NY.

5. Williamson, R. (1980) The processing of hnRNA and its relation to mRNA, in *Gene Expression: The Production of RNAs* (Goldstein, L. and Prescott, D. M., eds.), Academic, NY, pp. 547–562.

6. Cohen, S. and Stanley, W. (1942) The molecular size and shape of the nucleic acid of tobacco mosaic virus. *J Biol Chem* **144,** 589–598.

7. Noll, H. and Stutz, E. (1968) The use of sodium and lithium dodecyl sulfate in nucleic acid isolation. *Methods Enzymol.* **12,** Part B, 129–155.

8. Glisin, V., Crkvenjakov, R., and Byus, C. (1974) Ribonucleic acid isolated by cesium chloride centrifugation. *Biochemistry* **13,** 2633–2637.

9. Kirby, K. S. (1968) Isolation of nucleic acids with phenolic solutions. *Methods Enzymol.* **12,** part B, 87–99.

10. Ingle, J. and Burns, R. G. (1968) The loss of ribosomal ribonucleic acid during the preparation of nucleic acod from certain plant tissues by the detergent-phenol method. *Biochem. J.* **110,** 605,606.

11. Grinnan, E. and Mosher, W. (1951) Highly polymerized ribonucleic acid: preparation from liver and depolymerization. *J. Biol .Chem.* **191,** 719–726.

12. Cox, R. A. (1968) The use of guanidinium chloride in the isolation of nucleic acids. *Methods Enzymol.* **12,** Part B, 120–129.

13. Chirgwin, J. M., Przybyka, A. E., Mac Donald, R. J., Rutter, W. J. (1979) Isolation of biologically active ribonucleic acid from sources enriched in ribonuclease. *Biochemistry* **18,** 5294–5299.

14. Chomczynski, P. and Sacchi, N. (1987) Single-step method of RNA isolation by acid guanidinium thiocyanate-phenol-chloroform extraction. *Anal. Biochem.* **162,** 156–159.

15. Macfarlane, D. E. and Dahle, C. E. (1993) Isolating RNA from whole blood—the dawn of RNA-based diagnosis? *Nature* **362,** 186–188.

16. Dahle, C. E. and Macfarlane, D. E. (1993) Isolation of RNA from cells in culture using Catrimox-14 cationic surfactant. *BioTechniques* **15,** 1102–1105.

2

Large and Small Scale RNA Preparations from Eukaryotic Cells

Wolfgang Uckert, Wolfgang Walther, and Ulrike Stein

1. Introduction

A mammalian cell contains approx 10^{-5} µg of RNA which consists mainly of rRNA and in smaller amounts of a variety of low-mol-wt RNA species. These RNAs are of defined size and sequence. The ability to isolate clean, intact, and DNA-free RNA is a prerequisite in analyzing gene expression and cloning genes. The regulation of gene expression, e.g., the analyses of detailed function of transcription factors, promoter and enhancer sequences, and RNA synthesis and processing as well as the analyses of gene expression after transfer of genes of interest into eukaryotic cells are common areas of investigation in molecular biology. Important for the study of regulation of gene expression is the ability to isolate, analyze, and quantify RNA molecules, specifically mRNAs coding for proteins of interest. Furthermore, RNA is needed in order to copy it into double-stranded DNA for cloning and production of a cDNA library. The critical first step in the construction of a cDNA library is the efficient isolation of undegraded total RNA. The major difficulty in RNA isolation is the presence of ribonucleases found in virtually all tissues and liberated either during cell lysis or accidentally introduced in traces from other potential sources.

The isolation of total RNA from eukaryotic cells and the purification of mRNA has been described in detail in a variety of methods for laboratory application (1,2). These methods rely either on the use of guanidinium thiocyanate to disrupt the cells followed by a centrifugation in cesium chloride solutions to separate the RNA from other cellular components (3) or use guanidinium chloride which readily dissolves and biologically inactivates proteins (4). By this, all ordered secondary structure is lost while the secondary structure of deoxyribonucleic nucleic acids is not affected and remains in its native form. Further-

From: *Methods in Molecular Biology, Vol. 86: RNA Isolation and Characterization Protocols*
Edited by: R. Rapley and D. L. Manning © Humana Press Inc., Totowa, NJ

more, a number of methods have been developed to isolate RNA from mammalian cells grown in monolayer or suspension cultures *(2)*. However, these traditional RNA isolation methods are not useful in cases where multiple samples have to be processed, e.g., after gene transfer into recipient cells and subsequent analyses of gene expression in a large numbers of cell clones or for the investigation of great numbers of tissue samples. Traditional RNA isolation methods are also not suitable if individual samples are small because the sample is limited.

The RNA isolation protocol introduced here is based on a lithium chloride/urea method *(5)* and fulfills the following criteria: the procedure is very simple, it includes short incubation and reaction times, it needs relatively small amounts of cells or tissue, and it is suited for the processing of a great number of samples as a multiple-sample preparation. The method can be performed both on a large-scale but can easily be miniaturized to a RNA minipreparation protocol for small amounts of material from a variety of sources including mammalian cells, tissue samples, and cryo-sections. Ribonucleases are effectively inhibited by high salt and urea concentration. RNAs are selectively precipitated with lithium chloride while other components, such as DNA, polysaccharides, and proteins remain in solution. Some specific advantages of this method are: no usage of harmful chemicals (except phenol), no necessity of ultracentrifugation steps, the possibility to store samples after the lithium chloride/urea treatment without any degradation of RNA and therefore to accumulate a large number of probes, and samples are free of DNA without DNase treatment. Finally, the method is inexpensive in comparison to RNA isolation kits that are commercially available from different suppliers. RNA isolated according to the given protocols can be used both for Northern-blot analyses (**Fig. 1**) and RT-PCR experiments (**Fig. 2**).

2. Materials

To avoid any potential RNAse contamination, wear gloves for the preparation of all solutions.

1. Lithium chloride/urea solution: $3M$ LiCl and $6M$ urea dissolved in distilled and autoclaved DEPC (diethyl pyrocarbonate)-treated water [this is available as RNAse-free water from USB, (Cleveland, OH)], or prepare by adding 20 µL DEPC to 100 mL water and autoclave; *Caution: DEPC is suspected to be carcinogenic.* Filter the solution through a 0.45-µm disposable filter (it is then unnecessary to autoclave the solution). The solution is stable at 4°C for up to 6 mo.
2. TES buffer: 10 mM Tris-HCl, pH 7.6, 1 mM EDTA, and 0.5% SDS (sodium dodecyl sulfate) in distilled DEPC-treated water. The TES buffer has to be autoclaved and then stored at room temperature preventing precipitation of SDS.
3. Phenol/chloroform solution: For the preparation of the solution Tris-buffer saturated phenol *(Caution: Phenol is corrosive and toxic)* with pH 7.0 should be used

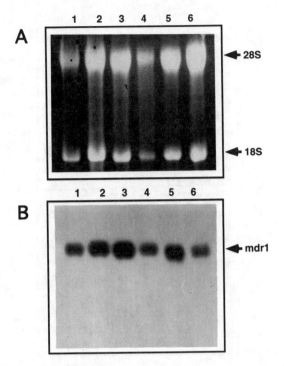

Fig. 1. Ethidium bromide stained agarose gel **(A)** and Northern-blot **(B)** of total cellular RNA isolated from human colon carcinoma cell lines KM12 (lane 1), HCT116 (lane 2), and HCT15 (lane 3); and from human osteosarcoma specimen (lanes 4–6). (A) RNAs were prepared by the LiCl/urea method (*see* **Subheading 3.1.**) and 10 µg of each RNA were subjected to gel electrophoresis in 1.2% agarose gel containing 6.7% formaldehyde (*see* Chapter 13). The arrows indicate the 28S and 18S bands that represent the characteristic eukaryotic ribosomal RNAs. (B) For the Northern-blot analyses, RNAs were transferred from the agarose gel onto nitrocellulose filters (Hybond N[+], Amersham) and hybridized to a multidrug resistance gene (mdr1) specific ^{32}P-labeled DNA-probe according to standard hybridization protocols *(2,6)*. For autoradiography, filters were exposed 24 h to X-ray film (Fuji). The bands on the autoradiograph represent the mdr1-specific transcripts (mRNAs), that are expressed at different levels in the cell lines or tumor tissues, respectively.

> *(2)*. Add to phenol an equal volume of chloroform (analytical grade; chloroform should contain 1/24 vol isoamyl alcohol), and store at 4°C.

4. Sodium acetate solution: Prepare a 3*M* sodium acetate solution in distilled DEPC-treated water and adjust to pH 7.0 with diluted acetic acid before adjusting with water to the final volume. Filter the solution through a 0.45-µm disposable filter, autoclave the solution, and store at 4°C.

5. 70% DEPC/ethanol: Prepare the 70% ethanol by mixing absolute ethanol with DEPC-treated distilled water, and store at –20°C.

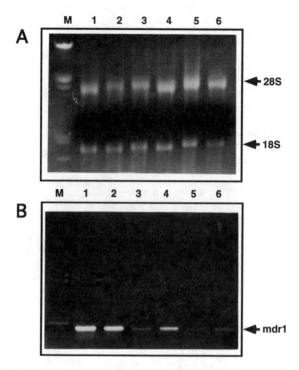

Fig. 2. Ethidium bromide stained agarose gels of minipreparation RNAs **(A)** and of Reverse Transcriptase PCR (RT-PCR) using minipreparation RNA **(B)** from human colon carcinoma cell lines HCT15 (lane 1), HCT116 (lane 2), KM12 (lane 3), and from cryo-sections of human sarcoma tissues (lanes 4–6); 1 kbp ladder (size markers, Bethesda Research Labs; lane M). (A) RNAs were prepared by the minipreparation protocol (*see* **Subheading 3.2.**) and 5 µg of each miniprep RNA were separated by gel electrophoresis in a 1.2% agarose gel containing 6.7% formaldehyde. (B) RT-PCR was carried out with 1 µg of each minipreparation RNA using the GeneAmp RNA PCR Kit (Perkin-Elmer). Conditions for the reaction were chosen according to the manufacturer. For the RT reaction, random hexamer primers were employed, and for the PCR mdr1-specific primers were used that yield a 0.167 kbp PCR-product *(7)*. The mdr1-specific primers are designed to bind to two separate exons of the mdr1 gene. Therefore, potential contaminations of RNA minipreparations with high-mol-wt DNA would create additional PCR-products of significantly larger sizes (<1 kbp). Thus, absence of these bands indicates the high-quality of the RNA minipreparations. After RT-PCR, the PCR-products were run in a 1.5% agarose gel and stained with ethidium bromide. The arrow indicates the mdr1 specific product.

6. Phosphate-buffered saline (PBS): 1% NaCl, 0.025% KCl, 0.14% Na_2HPO_4, 0.025% KH_2PO_4 (all w/v), pH 7.3. Make up the solution in distilled water, autoclave, and store at 4°C.

3. Methods

3.1. Isolation of Total Cellular RNA by LiCl/Urea Procedure

To ensure good-quality RNA preparations and to prevent RNA degradation, wearing gloves is an essential requirement for the whole preparation procedure. In addition, autoclave all tubes, tips, and vials that will be used during the RNA preparation.

1. Wash cells (after removal of cell culture medium) or tissue samples gently but quickly, two times with 5 mL ice-cold PBS. Then add 2–5 mL ice-cold LiCl/urea solution to the cells or tissue sample, transfer to a Dounce homogenizer and homogenize the probes by 10–20 strokes in the homogenizer (*see* **Note 1, Subheading 4.1.**). Make sure that the probes are kept on ice during homogenization.
2. After homogenization, transfer the samples to 16-mL polypropylene tubes (Nalgene, Rochester, NY) and let stand at 4°C overnight (*see* **Note 2**).
3. Spin the samples at 12,000 rpm for 20 min at 4°C, discard supernatants (*see* **Notes 1 and 2, Subheading 4.1.**), then add half of the original volume LiCl/urea solution and vortex thoroughly to dissolve the RNA pellet again (*see* **Note 3**).
4. Spin the samples at 18,000g for 20 min at 4°C, discard the supernatant and add half of the original volume (same vol as at **step 3**) of TES buffer to the RNA pellet. Vortex the sample as long as the pellet is almost readily dissolved in TES (*see* **Note 4**).
5. Add the same volume of phenol/chloroform to the samples as it was added in **step 4** and vortex thoroughly. At this point, the solution will become milky due to the water-insoluble organic solvents and the SDS precipitation.
6. Spin the samples at 12,000g for 10 min at 4°C and transfer the supernatant carefully to another 16-mL polypropylene tube using blue Eppendorf tips (*see* **Note 5**). Then add $^{1}/_{10}$ vol of 3M sodium acetate to the probes and mix well.
7. Add 2.5 vol ice-cold absolute ethanol to the samples and keep at –70°C or on dry ice for 30 min to precipitate the RNA (*see* **Note 6**).
8. Spin the frozen samples at 18,000g for 30 min at 4°C, discard the supernatant, wash the RNA pellets with 5 mL 70% DEPC/ethanol and spin again at 18,000g for 15 min at 4°C (*see* **Note 6**).
9. After removal of the 70% ethanol, dry the RNA pellets, and dissolve the RNA in 50–100 mL DEPC-treated water and determine RNA concentration at 260 nm in a spectrophotometer. Store the RNA samples at –20°C, where they are stable for months.

3.2. Minipreparation of Total Cellular RNA

This minipreparation protocol for the isolation of total cellular RNA is based on the previous LiCl/urea procedure and represents a shortened quick variation of this method in which the same materials are used (*see* **Subheading 2**). The protocol is useful if only very small amounts of cells or tissue samples are available for RNA isolation.

1. Employing this protocol, it is appropriate to use $1-5 \times 10^5$ cells or tissue cryo-sections (*see* **Note 7**). Wash the cells with 1 mL ice-cold PBS and add 100–500 mL ice-cold LiCl/urea solution to the cells or cryo-sections and transfer the samples to 1.5-mL Eppendorf tubes. Make sure that the cells are detached from the bottom of the dish, otherwise scrape the cells or remove by pipeting several times before transferring the samples to the tubes. Let the sample stand on ice for 20 min (for cryo-sections: *see* **Notes 7** and **8**).

2. Spin the samples at 15,800g in an Eppendorf centrifuge for 20 min at 4°C.

3. Discard the supernatants carefully and add half of the original vol of ice-cold LiCl/urea solution and vortex thoroughly to dissolve the pellet completely (*see* **Note 8**).

4. Spin the samples at 15,800g in an Eppendorf centrifuge for 20 min at 4°C and discard the supernatants.

5. Add half of the original vol TES buffer and dissolve the pellets by vortexing and/or pipeting (*see* **Note 9**). If the pellets have been dispersed, then add the same vol phenol/chloroform (room temperature) to the probes and vortex thoroughly.

6. Spin the samples at 11,600g for 10 min at 4°C and transfer the supernatants to another Eppendorf tube.

7. Add $\frac{1}{10}$ vol 3M sodium acetate solution and 2.5 vol absolute ethanol and precipitate RNA at -70°C or on dry ice for 20 min (*see* **Note 5**).

8. Spin the samples at 15,800g for 15 min at 4°C and wash the pellets once with 500 μL 70% DEPC/ethanol. Centrifuge again at 15,800g for 5 min 4°C, discard the ethanol and dry the pellets (*see* **Note 6**). Dissolve the RNAs in 10–20 μL DEPC-treated water and calculate the RNA yields by spectrophotometry at 260 nm (*see* Chapter 12).

4. Notes

4.1. LiCl/Urea-Procedure

1. For the usual preparation of total cellular RNA from cell cultures it is sufficient to use $5 \times 10^5-1 \times 10^6$ cells or tissue samples of 0.125–1.0 cm^3 (approx 50–200 mg tissue). The cell number/density of cell cultures or the size of the tissue sample determines the volume of the LiCl/urea solution in **Subheading 3.1., step 1**, Preparations from this amount of cells yields 50–150 μg, and from tissues 10–20 μg of total cellular RNA.

   The homogenization step is quite important to dissolve as much RNA as possible in LiCl/urea solution and to shear the high-mol-wt DNA. During this step, make sure that the homogenate is not too viscous, since this would be disadvantageous for pelleting the RNA in **Subheading 3.1., step 3**, of the procedure: the RNA pellet would not adhere properly to the tube wall (not dense enough) and could slip away during removal of supernatant. Therefore, if the homogenate is very viscous, then add 1 or 2 more mL of LiCl/urea solution after the homogenization and vortex thoroughly.

   It is noteworthy that at this stage samples (including RNA) are stable at 4°C over a period of several weeks. Thus, it is possible to collect samples for the preparation of several RNAs at one time without the danger of RNA degradation during storage.

2. If tissue is used for RNA preparations, the remaining tissue debris should also be transferred to the tube, since this tissue material will be removed by the phenol/chloroform treatment in **Subheading 3.1., step 5**, Try to use transparent or semi-transparent tubes which allows identification of the RNA pellets easily, because the pellets are almost transparent. These tubes should be phenol-resistant (such as polypropylene), so that it is not necessary to transfer the probes again before adding phenol/chloroform to the samples.

3. This step functions as a washing procedure to remove residual high-mol-wt DNA. For this reason it is essential to dissolve the whole RNA pellet, which will reappear after the second centrifugation step.

4. This step liberates the RNA preparation from cell debris and protein. The thorough vortexing of the samples ensures good-quality of RNA that is free of any protein contamination. This step should be carried out at room temperature to avoid precipitation of SDS.

5. The removal of the RNA-containing aqueous supernatant can sometimes be difficult, because this tends to be a viscous solution. To avoid the interphase coming off with the supernatant, keep the samples on ice after centrifugation and use tips for the removal of the supernatants, which have been cut at their top.

6. Take the frozen samples for the centrifugation; it is not necessary to let the samples warm up before spinning. At this point it is possible to interrupt the preparation—the frozen samples are stable for several weeks. Be careful when discarding the supernatants, since the RNA pellets can sometimes detach from the tube wall!

7. The homogenization of either the cells or the cryo-sections in LiCl/urea is not necessary using the minipreparation method. In this protocol the osmotic shock, caused by the high molarity of the LiCl/urea solution, is sufficient to disrupt the cell structures. Furthermore, if cell cultures are used for the RNA minipreparation (e.g., in 24- or 96-well cell culture dishes), the procedure at **steps 1–5** in **Subheading 3.2.** can be modified by using the following short-cut protocol: After washing the cells with 1 mL ice-cold PBS, 200 µL of LiCl/urea solution is added, and incubated for 5 min at room temperature. Then add $1/10$ vol sodium acetate and 2.5 vol absolute ethanol, mix the solution thoroughly and transfer to an Eppendorf tube and leave for 15 min at room temperature. Thereafter, centrifuge at 11,600g for 5 min at room temperature, discard the supernatants and continue as described for **Subheading 3.2., step 6**.
   If tissue cryo-sections are used for the RNA isolation, fresh sections (approx 0.5–cm in diameter) should be placed in 100–300 µL ice-cold LiCl/urea solution in an Eppendorf tube and left for 48–72 h at 4°C, vortex the probes during this incubation from time to time. This will help to dissolve as much RNA from the section as possible. Since the cryo-sections are not homogenized, they remain in the preparations till the phenol/chloroform treatment has been completed (**Subheading 3.2., step 5**). The RNA yields of the minipreparation range between 10–15 µg for cell cultures or 1–5 µg for cryo-sections.

8. At this stage it can be difficult to see the pellets. Therefore, discard the supernatants, so that a small amount of the LiCl/urea solution is left in the tube. This does not interfere with the subsequent preparation steps. If the amount of cells is very small or for tissue cryo-sections the **steps 3** and **4** in **Subheading 4.2.** can be skipped, since washing the pellets could lead to significant loss of RNA in the preparations.

9. At this point, it is essential to disperse the pellets thoroughly, since this determines the yield of RNA in the preparations: If vortexing is inefficient, use 100 μL Gilson-pipet in this step and pipet several times.

References

1. Janssen, K. (ed.) (1987) *Current Protocols in Molecular Biology*. Wiley & Sons, New York, pp. 4.0.1–4.2.8.

2. Sambrook, J., Fritsch, E. F., and Maniatis, T. (eds.) (1989) *Molecular Cloning: A Laboratory Manual*. Cold Spring Harbor Laboratory Press, Cold Spring Harbor, NY, pp. 7.3–7.25.

3. Glisin, V., Crkvenjakov, R., and Byus, C. (1974) Ribonucleic acid isolated by cesium chloride centrifugation. *Biochemistry* **13**, 2633–2637.

4. Cox, R. A. (1968) The use of guanidinium chloride in the isolation of nucleic acids. *Methods Enzymol.* **12**, 120–129.

5. Auffray, C. and Rougeon, F. (1980) Purification of mouse immunoglobin heavy-chain messenger RNAs from total melanoma tumor RNA. *Eur. J. Biochem.* **107**, 303–314.

6. Chen, C. J., Chin, J. E., Ueda, K., Clark, D. P., Pastan, I., Gottesman, M. M., and Roninson, I. B. (1986) Internal duplication and homology with bacterial transporter proteins in the mdr1 (P-glycoprotein) gene from multidrug-resistant cells. *Cell* **47**, 381–389.

7. Noonan, K. F., Beck, C., Holzmayer, T. A., Chin, J. E., Wunder, J. S., Andrulis, I. L., Gazdar, A. F., Willman, C. L., Griffith, B., von Hoff, D. D., and Roninson, I. B. (1990) Quantitative analysis of MDR1 (multidrug resistance) gene expression in human tumors by polymerase chain reaction. *Proc. Natl. Acad. Sci. USA* **87**, 7160–7164.

3

An Improved Rapid Method of Isolating RNA from Cultured Cells

David B. Batt, Gordon G. Carmichael, and Zhong Liu

1. Introduction

   The purification of good-quality RNA from tissue-cultured cells is essential for many applications and several methods exist for the isolation of total RNA. Most protocols rely on sodium dodecyl sulfate (SDS) *(1)* or guanidium thiocyanate *(2,3)* to simultaneously lyse cells and inactivate endogenous ribonucleases. In those procedures, the RNA is separated from cellular DNA and proteins by centrifugation through $5.7M$ CsCl, or by acid phenol extraction. In the former, the RNA passes through the $5.7M$ CsCl but proteins and large DNA molecules are excluded. In the latter method, protein and DNA partition into the acid phenol phase leaving the RNA in the aqueous phase. The method described below uses SDS to lyse cells and acidic phenol to remove DNA and proteins, and is a modification of the Stallcup and Washington procedure *(1)*. The modifications were originally made in order to reduce the amount of small DNA molecules, such as plasmids that are not efficiently removed in most other simple methods. Many researchers use transient transfection to study the fate of RNA produced from plasmid DNA. The removal of transfected plasmids, particularly if they have undergone replication, is critical in order to avoid hybrids, which can score as full-length RNA in S1 or ribonuclease protection assays (*see* Chapter 16).

2. Materials

1. Ice-cold phosphate-buffered saline (1X PBS): Store at 4°C.
2. Solution A: 10 mM EDTA (pH 8.0), 1% SDS. Store at room temperature.
3. 10 mg/mL proteinase K solution, in H_2O: Store at −20°C.
4. Solution B: 10 mM EDTA, pH 8.0, $0.1M$ sodium acetate, pH 4.0. Store at 4°C.

From: *Methods in Molecular Biology, Vol. 86: RNA Isolation and Characterization Protocols*
Edited by: R. Rapley and D. L. Manning © Humana Press Inc., Totowa, NJ

5. Solution C: Water-saturated phenol containing 0.04% (wt/wt) 8-hydroxyquinoline. Store at 4°C (*see* **Note 1**).
6. Solution D: Five parts solution C (phenol phase) mixed with one part chloroform/isoamyl alcohol (24:1). Store at 4°C with a layer of water above the organic layer.
7. 5*M* NaCl: Make this solution RNase-free by treatment with diethyl pyrocarbonate (DEPC): add DEPC to 0.1% (vol/vol) for 30 min. Autoclave for 30 min. DEPC is a suspected carcinogen and should be handled as such (*see* **Note 2**).
8. Ice-cold absolute ethanol: Store at –20°C.
9. 70% ethanol: Prepare this solution using DEPC-treated water.
10. DEPC-treated water: Treat water with 0.1% DEPC for 30 min followed by a 30 min autoclave treatment. DEPC is a suspected carcinogen and should be handled as such (*see* **Note 2**).

3. Method

The procedure given below is for isolating total RNA from a 100-cM^2 tissue culture dish. If smaller or larger dishes are used, the volumes of the solutions should be adjusted based on the surface area of the plates.

1. Rinse cells with ice-cold PBS.
2. Lyse cells by adding 2 mL of solution A to the plate. Collect lysate with a cell scraper into a 6-mL or 15-mL polypropylene tube.
3. Add 10 μL of 10 mg/mL proteinase K to the lysate, and incubate at 45–50°C for 30 min (*see* **Note 3**).
4. After proteinase K digestion add 2 mL solution B. Vortex briefly.
5. Add 4 mL solution C. Vortex for 10 s then incubate on ice for 15 min.
6. Centrifugation at 12,000*g* for 10 min at 4°C, remove the aqueous (top) phase, and place in a new tube (*see* **Note 4**).
7. To the aqueous phase add an equal volume of solution D. Vortex. Centrifuge for 10 min at 12,000*g*.
8. Mix the aqueous (top) phase (about 3.5. mL) with 130 μL of 5*M* NaCl and 2 vol of ice-cold ethanol (*see* **Note 5**).
9. Collect the RNA by centrifugation at 12,000*g* for 15 min at 4°C (*see* **Note 6**). Rinse the RNA pellet with 70% ethanol. Dry RNA pellet briefly, then resuspend in DEPC-treated water (*see* **Note 7**).

4. Notes

This procedure should yield about 100–200 μg of total RNA with an A$_{260/280}$ ratio of 1.7–1.8. This RNA can be used in ribonuclease protection assays, Northern blot analyses, and as a template for reverse transcription. Using this procedure, the first phenol extraction (solution C) removes approx 99% of the plasmid DNA. The subsequent extraction (solution D) removes about 60–70% of the remaining DNA. Additional extractions with solution D may be used if a further reduction in DNA is required for a specific application.

1. Solution C should be made with a molecular biology grade reagent and the phases allowed to separate before use.
2. The autoclave cycle is critical because it destroys the DEPC which can covalently modify RNA if not removed.
3. The proteinase K digestion is performed at 45–50°C. SDS inhibits nucleases and does not seem to interfere with the protein digestion.
4. After centrifugation in **Subheading 3.**, **step 6**, the DNA will partition to the interphase layer. It is critical to avoid the interphase when removing the aqueous layer.
5. It is important to precipitate the RNA with NaCl, as other salts, especially potassium salts, can precipitate SDS.
6. An incubation of 30 min at 0°C may somewhat increase the RNA yield.
7. It is important that the RNA pellet not be over-dried because such pellets can be very difficult to resuspend.

References

1. Stallcup, M. R. and Washington, L. D. (1983) Region-specific initiation of mouse mammary tumor virus RNA synthesis by endogenous RNA polymerase II in preparations of cell nuclei. *J. Biol. Chem.* **258(5),** 2802–2807.
2. Chirgwin, J. M., Przybyla, A. E., MacDonald, R. J., and Rutter, W. J. (1979) Isolation of biologically active ribonucleic acid from sources enriched in ribonuclease. *Biochemistry* **18(24),** 5294–5299.
3. Chomczynski, P. and Sacchi, N. (1987) Single-step method of RNA isolation by acid guanidinium thiocyanate-phenol-chloroform extraction. *Anal. Biochem.* **162(1),** 156–159.

4

Isolating RNA with the Cationic Surfactant, Catrimox-14

Christopher E. Dahle and Donald E. Macfarlane

1. Introduction

Cationic surfactants precipitate nucleic acids, presumably by forming reverse micelles in which the aliphatic tails of the surfactant face the aqueous phase, and the cationic head groups bind to the nucleic acid electrostatically. The precipitate can be dissolved in organic solvents, or it can be redissolved in water by the addition of salt. Selected cationic surfactants are capable of lysing cells, by solubilizing their protein and lipid components.

For reasons that are not clear, the efficiency of cell lysis and the efficiency of precipitation of RNA are independently influenced by the counterion *(1)*. The method we describe here uses Catrimox-14, an aqueous solution of $0.1M$ tetradecyltrimethylammonium oxalate, which lyses cells without difficulty, and precipitates RNA quantitatively *(2)*. Once precipitated in this way, RNA is protected from RNase. After centrifugation, DNA, the surfactant and other impurities are removed from the resulting pellet by extraction with a high concentration of lithium chloride, in which RNA is insoluble. The RNA is then dissolved from the pellet with water.

This method is designed for isolating RNA from small samples. It is particularly suitable for processing multiple samples each yielding about 10 μg RNA. Alternative methods for extracting the RNA from the pellet have been described, and may be superior for the detection of viruses in blood *(2)*.

2. Materials

1. Catrimox-14: $0.1M$ solution of tetradecyltrimethylammonium oxalate (Iowa Biotechnology, Coralville, IA).
2. $2M$ LiCl in RNase-free water.

From: *Methods in Molecular Biology, Vol. 86: RNA Isolation and Characterization Protocols*
Edited by: R. Rapley and D. L. Manning © Humana Press Inc., Totowa, NJ

3. 70% ethanol made with RNase-free water.
4. Microfuge, preferably variable speed.
5. Microcentrifuge tubes made of RNase-free, clear polypropylene or polyethylene.

3. Method

Note: All procedures should be performed while wearing gloves to reduce contamination of samples by RNase that are present on human skin. Plastic tips and centrifuge tubes straight from the packaging are usually RNase-free. Reusable equipment, like electrophoresis boxes, should be dedicated to RNA isolation.

1. Add 0.1 mL blood, about 1–10 million cells from suspension culture, or about 10 mg well-dispersed tissue, to 1 mL Catrimox-14 at room temperature. Mix well (*see* **Notes 1** and **2**).
2. After 10 min, centrifuge for 5 min at room temperature (*see* **Note 3**).
3. Aspirate and discard supernatant. Wash pellet with 1 mL RNase-free water. Avoid dislodging the pellet. Aspirate and discard the wash.
4. Add 0.5 mL 2*M* LiCl to the pellet. Vortex well (15 s), at least two times (*see* **Note 4**).
5. Centrifuge 5 min at room temperature at 10,000g. Discard supernatant (or save for DNA precipitation if necessary; *see* **Note 5**).
6. Wash pellet in 1 mL 70% ethanol. Centrifuge 5 min at room temperature at 10,000g.
7. Air-dry pellet 15–20 min.
8. Dissolve RNA in small volume of water or buffer (*see* **Notes 6–14**).

4. Notes

1. Rapid and complete mixing of the sample and the Catrimox-14 is important. There is no need to centrifuge cells from their culture medium.
2. For cells growing in monolayers, the Catrimox-14 can be applied directly to the cells *in situ*. Use sufficient Catrimox-14 to cover the surface of the cells. Collect the resulting suspension, and rinse the surface thoroughly with additional Catrimox-14, or with RNase-free water, to capture all of the precipitating RNA. Mix this suspension thoroughly before centrifugation.
3. The speed of the initial centrifugation from Catrimox-14 must be sufficient to precipitate the RNA, but must not over-compact the pellet. With blood samples or with 100,000 eukaryotic cells, we recommend using 10,000g. For samples of 1 million cells, use 1000g. When the sample has 50 million cells use 250g. Subsequent centrifugations from LiCl and ethanol washes are at full speed.
4. LiCl extracts DNA from the pellet better if it is dislodged and dispersed by vortex mixing prior to adding the LiCl.
5. When high cell loading is used, DNA released from the pellet will make the supernatant viscous. In this case, shear the DNA by drawing up the solution multiple times into a 1-mL disposable pipet tip, or with similar action through a 25-gage needle on a 1-mL syringe. It is generally advisable not to use high cell loading.

6. Be sure to dissolve RNA that adheres to the sides of the tube.
7. Heating RNA at 65°C for 10 min may help to dissolve RNA.
8. If there is a large amount of insoluble material in the final RNA preparation (as may happen with plant material and bacteria), remove by centrifuging after heating.
9. Use formaldehyde gels to check the integrity of small amounts of RNA. Dissolve the RNA in 5–10 μL of premix containing 10% (v/v) 10X MOPS, 17.5% (v/v) formaldehyde, 50% (v/v) formamide and RNase-free water. Heat at 65°C for 10 min, then add ⅕ volume of 50% (v/v) glycerol containing 1 mM EDTA, without tracking dyes.
10. If the starting material was less than 0.5 million cells, there may be insufficient material to assay by UV absorption (*see* Chapter 12, this volume).
11. The lithium chloride supernatant contains >98% of the genomic DNA. This can be recovered by precipitation with 2 vol of 100% ethanol added to the solution.
12. Successful isolation of undegraded RNA requires immediate mixing of the cells of the sample with the Catrimox-14. Frozen sections (20-μ thick) can be dropped into Catrimox-14. Tissue samples must be finely dispersed before adding to Catrimox-14.
13. RNA can be isolated directly from some Gram-negative bacteria. Add a suspension of a few mg of bacteria to 1 mL of Catrimox-14, and proceed as above. Some bacteria will not lyse in Catrimox-14, and may need to be pretreated with lysozyme or mechanical disruption.
14. Plant RNA has been successfully isolated from alfalfa seedlings. Chop 25–100 mg seedlings finely with a razor blade in the Catrimox-14 solution and allow debris to settle out. Transfer supernatant to new tube and continue as above.

References

1. Macfarlane, D. E. and Dahle, C. E. (1993) Isolating RNA from whole blood—the dawn of RNA-based diagnosis? *Nature* **362,** 186–188.
2. Dahle, C. E. and Macfarlane, D. E. (1993) Isolation of RNA from cells in culture using Catrimox-14 cationic surfactant. *BioTechniques* **15,** 1102–1105.

5

RNA Extraction from Formalin-Fixed and Paraffin-Embedded Tissues

Giorgio Stanta, Serena Bonin, and Rosella Perin

1. Introduction

Fixed and paraffin-embedded tissues from pathology department archives can be available for RNA expression analysis. We have already shown that RNA isolated from biopsy, surgical, or autopsy tissue, routinely processed by fixation and paraffin embedding, is not completely degraded. RNA fragments of around 100 bases in length or more are still present even in organs fixed at later stages after removal and very rich in RNase, such as pancreas *(1)*. Analysis of RNA from paraffin-embedded tissues allows us to use a large quantity of human tissues with any type of lesions collected in the pathology departments of any hospital. This method can be used to study the persistence in tissues of RNA virus genome or cell expression. Some procedures have been proposed to study this type of tissue *(1–3)*. Here is described a general method for RNA extraction from single 6–8 µm human tissue histological sections cut from paraffin blocks that can give constant and reproducible results.

RNA extraction from paraffin-embedded tissues is made up of four steps. The first one is the deparaffinization and hydration of the tissue sections. The second is the digestion with a proteolytic enzyme such as proteinase K in the presence of effective RNase inhibition to remove proteins from the samples. Then a phenol-H_2O/chloroform extraction followed by isopropanol precipitation is performed to obtain a sufficiently clean RNA. This can be used for successive analysis by reverse transcription to cDNA and PCR amplification.

2. Materials

1. All solvents described are purchased from commercial sources.
2. Digestion solution for RNA (100 mL): $4M$ guanidine thiocyanate (40 mL), $1M$ Tris-HCl, pH 7.6 (3 mL), sodium N-lauryl sarcosine 30% (2.4 mL), H_2O DEPC-

From: *Methods in Molecular Biology, Vol. 86: RNA Isolation and Characterization Protocols*
Edited by: R. Rapley and D. L. Manning © Humana Press Inc., Totowa, NJ

treated (54.6 mL). Before use add 0.28 μL of β-mercaptoethanol for each 100 μL of solution (*see* **Note 1**).

3. β-Mercaptoethanol: must be added to the digestion solution (0.28 μL every 100 μL of solution) just before use. Store it at 4°C.
4. Proteinase K 20 mg/mL: solubilize in 50% DEPC-treated water and 50% glycerol, and store at –20°C (*see* **Note 2**).
5. Phenol equilibrated in water: This is very sensitive to light and to oxidation. In order to reduce the rate of phenol oxidation, 8-hydroxyquinoline (0.1%) is added to the stock solution to retard this effect. The solution of phenol/water in use can be stored at 4°C in a dark bottle, but stock solutions must be stored at –20°C for up to 2 yr.
6. Glycogen: solution of 1.0 mg/mL in water. Store in aliquots at –20°C.
7. TE buffer (1X): 10 m*M* Tris-HCl, pH 8.0, 1 m*M* EDTA pH 8.0. (Stock solution 10X. Store at room temperature.
8. 50 m*M* Tris-HCl, pH 8.0. (Stock solution 1*M*.) Store at room temperature.
9. Ethanol for DNA precipitation (100 mL): ethanol 95% 96.7 mL with the addition of 3*M* sodium acetate pH 7.0, 3.4 mL, store at –20°C, ready to use.

3. Methods

3.1. Cutting Paraffin Sections

Sections can be cut (6–8 μm thick) with standard microtomes; there is no need to change the blade for every sample because paraffin itself is cleaning it at every cut. We clean the blade after each sample with xylene to eliminate paraffin residues. Place the cut sections in 1.5-mL tubes. To handle the sections more easily, it is better to obtain rolled up sections when they are cut (this may be obtained by decreasing the temperature of the paraffin blocks putting them in a freezer before cutting).

3.2. Deparaffinization

To have available tissues for the RNA extraction the first step is to eliminate the paraffin with solubilization in an organic solvent, such as xylene. Washing with ethanol is then necessary to discard completely the xylene that could interfere with the enzymes used in the successive steps.

1. Put a maximum of five sections of 6–8 μm per 1.5-mL Eppendorf tube and solubilize the paraffin twice with 1 mL of xylene for 5 min at room temperature. Centrifuge at high speed in a microcentrifuge for 10 min and save the pellet after each time. Note that at this step the pellet does not adhere very firmly to the bottom of the tube, so remove supernatant carefully without loss of tissue fragments.
2. Wash with 1 mL of absolute ethanol for 10 min and once with ethanol 95%, centrifuge in a microcentrifuge for 10 min and save the pellet after each time.
3. Discard the alcohol and air-dry the tissue with tubes open in a thermoblock at 37°C for approx 30 min.

3.3. Protein Digestion

To obtain purified RNA, the tissue proteins must be removed. The best method is to perform a digestion with a proteolytic enzyme.

1. Add to each dried sample 1 vol (100–300 µL on account of the quantity of tissue present in the sections) of digestion solution containing β-mercaptoethanol (0.28 µL every 100 µL of solution).
2. Add proteinase K to a final concentration of 6 mg/mL (43 µL of 20 mg/mL proteinase K every 100 µL of digestion solution).
3. Incubate overnight at 45°C with swirling.

3.4. RNA Extraction

1. Add to each tube 1 vol of phenol-water/chloroform in a 70:30 ratio.
2. Mix by vortexing, put in ice for 15 min, and then centrifuge (12,000g) for 20 min.
3. Save the upper aqueous phase avoiding the proteinaceous interface between the two phases and transfer it to a new tube (*see* **Note 3**).

3.5. RNA Precipitation

The final concentration of RNA from the solution can be made by precipitation with isopropanol using glycogen as carrier because of the low concentration.

1. Add 5 µL of glycogen 1 mg/mL and 1 vol of isopropanol to the aqueous phase and keep at –20°C for 48 h. A long time at –20°C is necessary to obtain an efficient precipitation of the fragmented RNA.
2. Centrifuge at 12,000g for 20 min at 4°C. RNA pellets do not adhere to the bottom of microcentrifuge tubes as firmly as DNA pellets; when decanting the supernatant keep the pellet in sight at all times.
3. Wash the pellet in 100 µL of ethanol 75%, kept at –20°C.
4. Centrifuge in a microcentrifuge for 5 min and air-dry the pellet.
5. Resuspend the RNA pellet in 25 µL of DEPC-treated water. Store at –80°C (*see* **Notes 4** and **5**).

3.6. Simultaneous DNA Extraction

It is possible to obtain also the DNA directly from the same sample used for the RNA extraction.

1. Equilibrate the bottom organic layer (phenol-chloroform) from the previous RNA extraction with one volume of Tris-HCl 50 m*M*, pH 8.0, at 4°C overnight.
2. Centrifuge in a microcentrifuge for 10 min at high speed and remove the upper aqueous layer.
3. Add 5 µL of glycogen 1.0 mg/mL and 3 vol of ethanol for DNA precipitation, store at –20°C overnight or over week-end.
4. Centrifuge in a microcentrifuge for 15 min at high speed to recover the pellet, wash with ethanol 75%, air-dry and resuspended in 25 µL of TE buffer.

4. Notes

1. Note that guanidine thiocyanate is toxic, and should be prepared in a chemical hood. Store the solution in a dark bottle because of light sensitivity. The solution can be stored at room temperature for up to 2 mo.
2. The high concentration of proteinase K is needed because of the presence of $1M$ guanidine thiocyanate that inhibits most enzymes but proteinase K *(5)*.
3. The proteinase K and the proteolysis residues must be eliminated to avoid interference with the successive steps. With the extraction of RNA with phenol/chloroform the proteins are localized at the interface between the organic and the upper aqueous phase while the RNA remains in the aqueous upper phase.
4. RNA extracted with this method from paraffin-embedded tissues is highly degraded with fragments ranging from 100 to 200 bases, the level of degradation is variable from sample to sample depending from the fixation and paraffin-embedding conditions. Usually formalin-fixed tissues give a high yield of RNA that can be used for further analysis. For an efficient RT-PCR analysis it is advisable to amplify fragments of 100 bases or less.
5. Sometimes as a result of excessive degradation of the RNA insufficient RNA is obtained from the extraction. To test the quality of the RNA preparation we suggest RT/PCR on an abundant mRNA like β-actin or GAPDH.

References

1. Stanta, G., and Schneider, C. (1991) RNA extracted from paraffin-embedded human tissues is amenable to analysis by PCR amplification. *BioTechniques* **11**, 304–308.
2. Weizsäcker, F. V., Labeit, S., Koch, H. K., Oehlert, W., Gerok, W., and Blum, H. E. (1991) A simple and rapid method for the detection of RNA in formalin-fixed, paraffin-embedded tissues by PCR amplification. *Biochem. Biophys. Res. Commun.* **174**, 176–180.
3. Godec, M. S., Asher, D. M., Swoveland, P. T., Eldadah, Z. A., Feinstone, S. M., Goldfarb, L. G., Gibbs, C. J., Jr., and Gajdusek, D. C. (1990) Detection of measles virus genomic sequences in SSPE brain tissue by polimerase chain reaction. *J. Med. Vir.* **30**, 237–244.
4. Gramlik, T. L., Fritsch, C., Shear, S., Sgoutas, D., Tuten, T., and Gansler, T. (1993) Analysis of epidermal growth factor receptor gene expression in stained smears and formalin-fixed, paraffin-embedded cell pellets by reverse transcription intron differential polymerase chain reaction. *Int. Acad. Cytol. Histol.* **15**, 317–322
5. Fisher, J. A. (1988) Activity of Proteinase K and RNase in guanidinium thiocyanate. *FASEB J.* **2**, A1126.

6

Extraction and Purification of RNA from Plant Tissue Enriched in Polysaccharides

Shu-Hua Cheng and Jeffrey R. Seemann

1. Introduction

The isolation of uncontaminated, intact RNA is essential for analyzing gene expression and for cloning genes. However, plant tissue is notorious for being a difficult source from which to isolate high-quality RNA with good yield. This difficulty is primarily due to the presence of naturally occurring polysaccharides and/or polyphenols that are released during cell disruption. These compounds form complexes with nucleic acids during tissue extraction and coprecipitate during subsequent alcohol precipitation steps (1–5). Depending on the nature and the quantity of these contaminants, the resulting alcohol precipitates can be gelatinous and difficult to dissolve. An RNA solution contaminated with polysaccharides and/or polyphenols is viscous and absorbs strongly at 230 nm. This UV absorption prevents an accurate quantitation of RNA concentration by a measurement of A_{260} (see Chapter 12). Furthermore, the contaminated RNA is not suitable for cDNA synthesis, reverse transcription PCR amplification, in vitro translation, or Northern analysis (6,7). The large number of publications of RNA isolation procedures reflects these difficulties and further demonstrates that the conditions required for successful isolation of RNA can differ significantly both between species and for the same species when grown under different environmental conditions (1–7,11). We are interested in the effects of rising atmospheric CO_2 on plant gene expression. Exposure of plants to elevated CO_2 (e.g., 100 ppm) often results in a several-fold increase of total leaf carbohydrates, particularly starch and soluble polysaccharides, relative to ambient (360 ppm) grown plants. Therefore, the problem of carbohydrate contamination during RNA isolation is generally even more pronounced in plants grown at elevated CO_2.

From: Methods in Molecular Biology, Vol. 86: RNA Isolation and Characterization Protocols
Edited by: R. Rapley and D. L. Manning © Humana Press Inc., Totowa, NJ

To isolate RNA, plant tissue is generally extracted with SDS/phenol *(2)* or guanidinium thiocyanate *(10)*. To overcome the problems of contaminating polyphenols and/or polysaccharides, techniques such as sedimentation in cesium chloride gradients *(11)* or differential solvent precipitation *(11)* have been employed. However, these techniques are not always effective in removing the contaminants *(9,10)*. We thus sought to develop a simple, reliable, and inexpensive method to isolate clean RNA with high yield. Our success in the isolation of leaf RNA from a variety of plant species grown at high CO_2 demonstrates the general applicability of this method *(8, see* **Table 1**). The protocol described here utilizes guanidinium thiocyanate and is modified from that reported by Chomczynski and Sacchi *(8)*. Guanidinium thiocyanate is effective for lysing cells and denaturing proteins, and when used in extraction buffers creates an immediate RNase-free environment. After tissue extraction, the homogenates are centrifuged at a moderate *g*-force to remove insoluble polysaccharides. The supernatant is then extracted using acid phenol/chloroform: RNA partitions to the aqueous phase whereas DNA and proteins are present in the interphase and the phenol phase. Most polysaccharides that remain in the aqueous phase are then selectively precipitated by potassium acetate *(11)*, and the RNA is further purified from residual contaminants by lithium chloride precipitation.

2. Materials

2.1. Solutions

1. 0.75*M* sodium citrate, pH 7.0 (with citric acid). 0.1% Diethyl pyrocarbonate (DEPC)-treated and autoclaved.
2. 10% (w/v) *N*-lauroylsarcosine.
3. Guanidinium thiocyanate buffer: Dissolve 250 g guanidinium thiocyanate in the manufacturer's bottle (without weighing) with 293 mL sterile deionized water, 17.6 mL 0.75*M* sodium citrate buffer, and 26.4 mL 10% *N*-lauroylsarcosine at 65°C (*see* **Note 1**).
4. Extraction buffer: Add 36 µL of β-mercaptoethanol per 5 mL of guanidinium thiocyanate buffer just before use. Polyvinylpolypyrrolidone (insoluble, 20% w/w) can be added to the extraction buffer if the tissue to be extracted contains a substantial level of polyphenols.
5. Chloroform/isoamyl alcohol 49:1 (v/v).
6. 2*M* sodium acetate, pH 4.0 (with acetic acid), 0.1% DEPC-treated and autoclaved.
7. Acid phenol (*see* **Note 2**).
8. 100% Isopropanol.
9. 70 and 100% Ethanol.
10. Deionized H_2O, 0.1% DEPC-treated and autoclaved.
11. 2*M* potassium acetate, pH 4.8, 0.1% DEPC-treated and autoclaved (*see* **Note 3**).
12. 10*M* LiCl, 0.1% DEPC-treated and autoclaved.

13. TNE buffer: 10 mM Tris-HCl, pH 7.5, room temperature, 150 mM NaCl, 1 mM EDTA.
14. TE buffer: 10mM Tris-HCl, pH 7.5, room temperature, 1mM EDTA.

2.2. Equipment

1. Mortar and pestle.
2. Liquid nitrogen.
3. 10-mL Nalgene Oak Ridge polypropylene tube, autoclaved.
4. Preparative centrifuge and microfuge.
5. Glass Pasteur pipet, baked at 200°C for at least 3 h.
6. Vortex.
7. 30-mL Nalgene Oak Ridge polypropylene tubes, autoclaved.
8. 15-mL Falcon snap-capped polypropylene tubes.
9. 1.5-mL microfuge tubes.

3. Methods

1. Grind 0.4 g of tissue in a mortar and pestle to a fine powder in liquid nitrogen. Do not allow the tissue to thaw.
2. Add 3.5. mL of extraction buffer and grind thoroughly while the tissue thaws. Transfer the homogenate to a 10-mL Oak Ridge tube. Rinse the mortar and pestle with 1 mL of extraction buffer and then transfer to the tube.
3. Centrifuge at 23,000g for 20 min at 4°C (swinging bucket rotor) (*see* **Note 4**).
4. Transfer the supernatant to 10-mL Oak Ridge tube using a baked glass Pasteur pipet (*see* **Note 5**).
5. Add 0.4 mL of 2M sodium acetate and mix by vortexing. Add 4 mL of acid phenol; vortex. Add 0.8 mL of chloroform/isoamyl alcohol; vortex.
6. Incubate the tube on ice for 15 min.
7. Centrifuge as in **step 3**.
8. Transfer the supernatant to a 30-mL Oak Ridge tube using a baked glass Pasteur pipet and add an equal volume of 2M potassium acetate; mix by vortexing.
9. Incubate the tube on ice for at least 30 min (*see* **Note 6**).
10. Centrifuge at 44,000g for 20 min at 4°C (fixed angle rotor) (*see* **Note 7**).
11. Transfer the supernatant to a 15-mL Falcon tube and add 0.6–1 vol of 100%, ice-cold isopropanol; mix by vortexing. Incubate the tube at –20°C for 45 min (*see* **Note 8**).
12. Centrifuge at 2700g for 20 min at 4°C (swinging bucket rotor).
13. Decant the supernatant. Wash the pellet with 1 mL of 70% ethanol; spin as in step 12 for 3 min and pour off as much of the ethanol as possible.
14. Dissolve the pellet in 400 µL of DEPC-treated water and transfer to a 1.5-mL microfuge tube.
15. Add 100 µL of 10M LiCl, and incubate at 4°C for at least 2 h.
16. Spin in a microfuge at 12,000g for 20 min at 4°C (*see* **Note 9**).
17. Wash the pellet twice in 1 mL of 70% ethanol.
18. Resuspend the pellet in 200 µL of TNE and add 500 µL of 100% ethanol. Incubate at –20°C for at least 15 min.

19. Spin the tube as in **step 16** for 5 min.
20. Wash the pellet twice with 1 mL of 70% ethanol.
21. Resuspend the pellet in 200–400 μL of TE, depending on the size of the pellet.
22. Make a 30–50-fold dilution of each sample and measure the absorbance at 230, 260, and 280 nm (*see* **Note 10**).

4. Notes

1. Since guanidinium thiocyanate is hazardous, it is best to prepare this solution in the manufacturer's bottle without weighing to minimize handling. When making a smaller quantity of the solution, wear gloves when weighing. This solution can be stored at least 3 mo at room temperature.
2. To minimize handling, dissolve 500 g crystal phenol in the manufacturer's bottle with 500 mL sterile deionized water. Store in 50-mL aliquots in a −20°C freezer. For routine use, this solution can be kept at 4°C for up to 1 mo.
3. Dissolve 19.63 g potassium acetate in 25 mL of deionized water and add glacial acetic acid until the pH is 4.8. Make up to 100 mL.
4. This initial spin removes the majority of insoluble polysaccharides, such as starch grains, by pelleting. The tissue debris forms a dark green pellet (if leaves were used) in the bottom of the tube and insoluble polysaccharides form a whitish gel-like layer on top of the tissue debris.
5. When pipeting the supernatant, care should be taken not to disturb the gel-like pellet since it is very soft. The volume of the supernatant should be approx 4 mL. If there is a significant volume loss due to a large pellet, then compensate with extraction buffer. In our experience, there can be up to a 1 mL loss of volume in tissue extracts that contain a large quantity of starch.
6. A longer incubation period (up to 30 min) improves the quality of RNA, particularly when there is a problem of protein contamination.
7. Polysaccharides form a whitish gel-like pellet. If the tissue used contains a high level of polysaccharides, a longer incubation period (up to 60 min) would help to precipitate more polysaccharides.
8. On complete mixing of isopropanol, no precipitate should be visible. Any visible precipitate indicates the presence of a significant quantity of polysaccharides. In our experience, incubation longer than 60 min results in precipitation of polysaccharides.
9. The RNA pellet should be white. The presence of an off-white, gel-like pellet indicates contamination by polysaccharides. In such an event, resuspend the pellet in 200 μL TE, add one volume of 2*M* potassium acetate, and incubate on ice for 30 min. Spin the 1.5-mL microfuge tube at 12,000g at 4°C for 20 min. Transfer the supernatant to a new 1.5-mL microfuge tube, and precipitate the RNA with 2.5 vol of 100% ethanol at −20°C for 15 min. Proceed to **step 16** in **Subheading 3**.
10. The success of an RNA isolation procedure may be judged by the quantity, quality, and integrity of RNA recovered. The RNA quality and quantity can be evaluated by measuring spectrophotometric absorbance at 230, 260, and 280 nm. An $A_{260}:A_{230}$ ratio lower than 2 indicates contamination with polysaccharides and/or polyphenols, and an $A_{260}:A_{280}$ ratio below 1.7 indicates contamination with pro-

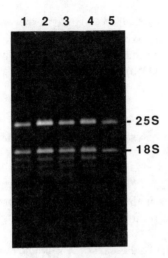

Fig. 1. Electrophoretic analysis of RNA isolated from various species. Two micrograms of total RNA isolated from leaves of (1) *Ajuga reptans*, (2) *Petroselinum hortense,* (3) *Plantago lanceolata*, (4) *Antirrhiinum majus*, (5) *Nicotiana sylvestris* were electrophoresed on a nondenaturing agarose gel (1.4% agarose in TBE buffer containing 0.5 µg/mL ethidium bromide).

Table 1
Qualitative and Quantitative Evaluation of Isolated RNA

| Species | Tissue | Absorbance ratio | | Yield |
		A_{260}/A_{230}	A_{260}/A_{280}	µg RNA/g FW
Ajuga reptans (common bugle)	leaf	2.02	2.14	509
Antirrhiinum majus (snapdragon)	leaf	2.23	2.22	929
Arabidopsis thaliana	leaf	2.33	1.98	9.15
Petroselinum hortense (parsley)	leaf	2.20	2.13	651
Phaseolus vulgaris	leaf	2.30	1.96	8.08
Plantago lanceolata (plaintain)	leaf	2.17	2.19	642
Spinacia oleracea (spinach)	leaf	2.20	2.10	733
Nicotiana sylvestris (tobacco)	leaf	2.10	1.90	607
Nicotiana sylvestris (tobacco)	root	2.35	2.08	525

teins (*see* Chapter 12). The integrity of RNA can be assessed by the intactness of the 25S and 18S ribosomal RNA bands in an agarose gel (*see* Chapter 13).

11. By using this protocol, we have successfully isolated highly purified RNA with high yield from a variety of plants. The RNA preparation is free of DNA (a common problem with many other protocols, **Fig. 1**). There was no apparent degradation of RNA as judged by the clarity and intactness of ribosomal RNA bands (**Fig. 1**). Also, the values of A_{260}:A_{230} and A_{260}:A_{280} typically were about 2, indicating little or no contamination of protein and polysaccharides (**Table 1**). The

average RNA yield is greater than 500 µg/g fresh weight of tissue (**Table 1**). This amount allows numerous experiments. This method is easily scaled up or down, and RNA prepared by this method is suitable for poly(A$^+$) selection (*see* Chapter 11), Northern analysis, cDNA synthesis or RT-PCR amplification *(8)*.

References

1. Lopez-Gomez, R. and Gomez-Lim, M. A. (1992) A method for extracting intact RNA from fruits rich in polysaccharides using ripe Mango mesocarp. *HortScience* **27,** 440–442.
2. Mitra, D. and Kootstra, A. (1993) Isolation of RNA from apple skin. *Plant Mol. Biol. Reptr.* **11,** 326–332.
3. Newbury, H. J. and Possingham, J. V. (1977). Factors affecting the extraction of intact ribonucleic acid from plant tissues containing interfering phenolic compounds. *Plant Physiol.* **60,** 543–547.
4. Schultz, D. J., Craig, R., Cox-Foster, D. L., Mumma, R. O., and Medford, J. I. (1994) RNA isolation from recalcitrant plant tissue. *Plant Mol. Biol. Reptr.* **12,** 310–316.
5. Wang, C.-S. and Vodkin L. O. (1994) Extraction of RNA from tissues containing high levels of procyanidins that bind RNA. *Plant Mol. Biol. Reptr.* **12,** 132–145.
6. Lay-Yee, M., DellaPenna, D., and Ross, G. S. (1990) Changes in mRNA and protein during ripening of apple fruit (*Malus domestica* Borkh. cv. Golden Delicious). *Plant Physiol.* **94,** 850–853.
7. Tesniere, C. and Vayda, M. E. (1991) Method for the isolation of high-quality RNA from grape berry tissues without contaminating tannins or carbohydrates. *Plant Mol. Biol. Reptr.* **9,** 242–251.
8. Chomczynski, P. and Sacchi, N. (1987) Single-step methods of RNA isolation by acid guanidinium thiocyanate-phenol-chloroform extraction. *Anal. Biochem.* **162,** 156–159.
9. Glisin, V., Crkvenjakov, R., and Byus, C. (1974) Ribonucleic acid isolated by cesium chloride centrifugation. *Biochemistry* **13,** 2633–2637.
10. Manning, K. (1991) Isolation of nucleic acids from plants by differential solvent precipitation. *Anal. Biochem.* **195,** 45–50.
11. Ainsworth, C. (1994) Isolation of RNA from floral tissue of *Rumex acetosa* (sorrel). *Plant Mol. Biol. Reptr.* **12,** 198–203.

7

Isolation of Plant Mitochondrial RNA from Green Leaves

Fei Ye, Wolfgang O. Abel, and Ralf Reski

1. Introduction

In plant cells, mitochondrial RNA (mtRNA) constitutes about only 1% of the total RNA. From this, most are ribosomal RNAs. Thus, isolation of high-purified mtRNA is necessary not only for construction of a mitochondrial cDNA library, but also for the analysis of plant mitochondrial transcription. Several methods have been frequently used for isolation of plant mtRNA (1–3). However, these mtRNA preparations may be heavily contaminated by chloroplast RNA (cpRNA), especially when mtRNA is isolated from green leaves (1,4). It is believed that the cpRNA sticks to the mitochondrial membrane and therefore persists after gradient purification of mitochondria. Although micrococcal nuclease would be the enzyme to remove the non-mtRNA from mitochondrial membranes prior to lysis of mitochondra, treatments with micrococcal nuclease for the mtRNA isolation from green leaves have not been effective (4).

We report here a modified procedure of mtRNA isolation based on the combination of RNase A/guanidine thiocyanate/CsCl centrifugation. In our procedure, mitochondria are first separated from other subcellular components, such as nuclei and plastids by differential centrifugation of leaf homogenates. The crude mitochondria are further purified by sucrose gradient centrifugation. To eliminate cpRNA, the purified mitochondria are treated with RNase A. Subsequently, RNase A is inactivated and mitochondria are lysed by adding guanidine thiocyanate in high concentration. As a strong protein denaturant, guanidine thiocyanate can inactivate nucleases very efficiently (5). mtRNA is pelleted through a CsCl gradient. Finally, coprecipitated single-stranded DNA in the CsCl gradient can be removed from mtRNA by LiCl precipitation (6).

From: *Methods in Molecular Biology, Vol. 86: RNA Isolation and Characterization Protocols*
Edited by: R. Rapley and D. L. Manning © Humana Press Inc., Totowa, NJ

2. Materials (*see* Notes 1 **and** 2)

1. 5.7M CsCl solution: 10 mM EDTA, pH 7.5, DEPC-treated.
2. Denaturation buffer: 50% formamide, 12% formaldehyde, 1X MOPS buffer (40 mM MOPS, 10 mM sodium acetate, 1 mM Na$_2$-EDTA, pH 7.0), freshly mixed before use.
3. 1 mg/mL Ethidium bromide: DEPC-treated, storage at –20°C.
4. Extraction buffer: 0.35M sorbitol, 50 mM Tris-HCl, pH 8.0, 5 mM EDTA, 0.1% BSA, 0.25-mg/mL each spermine and spermidine, storage at 4°C, then add β-mercaptoethanol to 0.2% (final concentration) just before use.
5. 4M Guanidinium thiocyanate: in 100 mM Tris-HCl, pH 7.5 (storage at 4°C), add β-mercaptoethanol to 1% (final concentration) just before use. Storage at 4°C.
6. 7.5M Guanidinium-HCl: 10 mM DTT, pH 7.5 (adjusted with NaOH), filtrate, storage at 4°C.
7. 2M and 4M LiCl: DEPC-treated, storage at 4°C.
8. Loading buffer: 50% glycerol, 0.25% bromophenol blue, 1 mM EDTA, DEPC-treated, storage at –20°C.
9. 10X MOPS buffer: 0.4M MOPS, 0.1M sodium acetate, 10 mM Na$_2$-EDTA, pH 7.0, DEPC-treated.
10. 2M potassium acetate: pH 5.5, DEPC-treated.
11. 2M Sodium acetate: pH 7.0, DEPC-treated.
12. 5% Sodium lauryl sarcosinate
13. TE buffer: 10 mM Tris-HCl, pH 8.0, 1 mM EDTA, DEPC-treated.
14. Wash buffer: 350 mM sorbitol, 50 mM Tris-HCl, pH 8.0, 20 mM EDTA.

3. Methods

3.1. Isolation of Mitochondria

All steps must be carried out at 4°C in a cold room. Solutions, bottles, and so on, should be kept in wet ice.

1. Harvest 20 g of 4-6-wk-old fresh green leaves from rapeseed or other plants, cut into small segments, and chill in 200 mL ice-cold extraction buffer.
2. Homogenize leaf tissue in a Waring blender at high speed three times (each time 5 s with 10 s breaks in between). Filter the homogenate through two layers of Miracloth into 250-mL cold centrifuge bottles.
3. Centrifuge the filtrate at 2000g for 10 min in a swing out rotor. Carefully transfer the supernatant to new bottles and centrifuge at 10,000g for 20 min in a swing out rotor.
4. Resuspend pellet in 100 mL extraction buffer and repeat **step 3** once again.
5. Resuspend the mitochondrial pellet in 20 mL ice-cold wash buffer.
6. Carefully layer each 10 mL mitochondrial suspension on top of a sucrose step gradient (9 mL 0.9M/11 mL 1.5M/9 mL 1.75M in wash buffer) and centrifuge for 60 min at 80,000g in a swing out rotor (*see* **Note 3**). Collect the mitochondria from the 0.9M/1.5M sucrose interface (yellow band) with wide-bore pipets and then dilute with 5 vol of wash buffer over a period of 15–20 min (*see* **Note 4**).

7. Pellet the mitochondria by centrifugation at 10,000g for 20 min in a swing out rotor and resuspend in 1 mL ice-cold wash buffer.

3.2. Isolation of mtRNA

1. Coincubated mitochondria with 20 µg/mL RNase A for 60 min on the ice.
2. Add 5 vol of 4M guanidine thiocyanate solution to the mitochondria, then add 0.5 vol 5% sodium lauryl sarcosinate after 60 s at room temperature and mix by vortexing. Centrifuge the mixture at 5000g for 5 min in a swing out rotor to remove insoluble debris.
3. Layer each 3.2. mL mixture onto a 1.1. mL cushion of DEPC-treated 5.7M CsCl solution.
4. Carry out ultracentrifugation at 22,000g for 14 h in a swingout rotor.
5. Carefully aspirate the supernatant solution and cut off the top part of the centrifuge tube that was in contact with the homogenate (all steps should avoid contamination of finger RNase).
6. Dissolve the RNA pellet by extensive vortexing in 1 mL 7.5M guanidinium-HCl solution.
7. Add 0.05 vol of 2M potassium acetate pH (5.5) and 0.5 vol of ethanol to the mixture.
8. Incubate at –20°C for 4 h and precipitate the mtRNA at 5000g for 10 min in a swingout rotor.
9. Precipitate the recovered mtRNA by adding 0.1 vol of 2M sodium acetate (pH 7.0) and 2.5 vol of ethanol. Store at –20°C overnight. Centrifuge for 30 min at 10,000g.
10. Wash the mtRNA pellet with 70% ethanol, vacuum dry, and dissolve in 0.5 mL TE.
11. To obtain mtRNA free from single-strand DNA, add an equal volume of 4M LiCl to dissolved RNA, incubate at 4°C overnight, and collect the mtRNA by centrifugation at 10,000g for 20 min in a microcentrifuge. Wash once with 2M LiCl and two times with 70% ethanol.
12. Vacuum dry the mtRNA and dissolve in 50 µL DEPC-treated water. Estimate the yield of mtRNA by measuring the absorbance at 260 nm (*see* Chapter 12).
13. For long-term storage, then add 0.3 vol of sodium acetate and 2.5 vol of ethanol to the mtRNA, and store at –70°C. Precipitate the RNA just before use.

This mtRNA preparation procedure will yield 0.3–0.5 µg mtRNA per gram fresh leaves. As a control, MS2 phage RNA (Boehringer, Mannheim) may be treated using the same conditions as mtRNA isolation. We have tested that the isolated RNA is intact (**Fig. 1**) and the contaminating cpRNA is totally eliminated (**Fig. 2**).

3.3. Formaldehyde Gel Electrophoresis (see Notes 6 and 7)

1. Melt 3.75 g of agarose in 220 mL DEPC-H$_2$O plus 30 mL 10X MOPS buffer. After the agarose is cooled to 60°C, add 50 mL of 37% (12.3.M) formaldehyde solution and pour the gel.

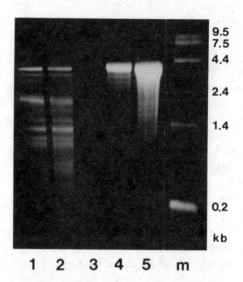

Fig. 1. Analysis of mtRNA preparation by electrophoresis in a 1.25% agarose-6% formaldehyde gel. Lane 1: mtRNA (5 μg) isolated with RNase treatment. Lane 2: mtRNA (5 μg) isolated without RNase treatment. Lane 3: MS2 phage RNA (10 μg) with RNase treatment. Lane 4: MS2 phage RNA (10 μg) without RNase treatment but using the conditions of mtRNA isolation. Lane 5: 10 μg MS2 phage RNA was directly loaded onto the gel. Lane m: RNA-ladder (BRL).

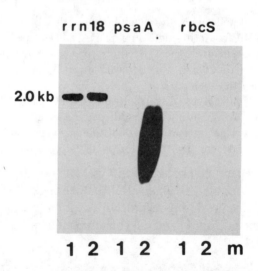

Fig. 2. Northern blot analysis of plant mtRNA preparations. The blots were probed with mitochondrial specific (*rrn*18), chloroplast-specific (*psa*A), and nucleus-specific (*rbc*S) gene probes. Lane 1: mtRNA (5 μg) isolated with RNase treatment. Lane 2: mtRNA (5 μg) isolated without RNase treatment. Lane m: RNA ladder (BRL).

2. Denature 9 μL RNA (5–10 μg) by adding an equal vol of RNA denaturation buffer, and incubate at 65°C for 5 min.
3. Add 1 μL of 1 mg/mL ethidium bromide, incubate at 65°C for an other 5 min, and place samples on ice for 5 min.
4. Add 2 μL of loading buffer to each sample and load on the prepared gel.
5. Carry out electrophoresis at 5 V/cm for 3–4 h. Soak the gel in DEPC-treated H₂O for 20 min to remove the formaldehyde, and photograph the gel.
6. Further Northern blot analysis can be carried out according standard methods *(7)* (*see* Chapters 15 and 16).

4. Notes

1. Wear disposable plastic or latex gloves during work. All reagents should be used for RNA work only and kept free of ribonuclease. All glassware should be baked at 160°C for 4 h. All plasticware to be used after the step of guanidinium thiocyanate treatment should be soaked in 0.2% DEPC for 12 h and autoclaved. All solutions to be used after the step of guanidinium thiocyanate treatment should be treated with 0.2% DEPC for 12 h and autoclaved.
2. DEPC is a carcinogen. DEPC-treatment of solutions and plasticware should be done in a chemical hood.
3. For sucrose gradient centrifugation, the prepared gradient should be allowed to equilibrate at 4°C overnight. After gradient centrifugation, wash buffer should be slowly added to collected mitochondria (over 15–20 min), this can minimize the osmotic shock.
4. For CsCl centrifugation, when different ultracentrifuge rotors are used, pay attention to maximum rotor speed and maximum run time.
5. The bulk of the DNA is removed using CsCl centrifugation. Since single-stranded DNA coprecipitates with the RNA in the CsCl gradient, 2*M* LiCl precipitations are necessary to obtain pure RNA preparations.
6. Formaldehyde is very toxic. Preparation and running of formaldehyde gels should be done in a chemical hood.
7. Reagent-grade formamide can be used directly. However, if any yellow color is present, formamide should be deionized by stirring it for 1 h with 5% (w/v) resin 501-X8 (D) (Bio-Rad). After filtration through Whatman No. 1 paper, deionized formamide should be stored in small aliquots at −70°C.

References

1. Stern, D. B. and Newton, K. J. (1986) Isolation of plant mitochondrial RNA. *Methods Enzymol.* **118,** 488–496.
2. Schuster, A. M. and Sisco, P. H. (1986) Isolation and characterization of single-stranded and double-stranded RNAs in mitochondria. *Methods Enzymol.* **118,** 497–507.
3. Schuster, W., Hiesel, R., Wissinger, B., Schobel, W., and Brennicke, A. (1988) Isolation and analysis of plant mitochondria and their genomes, in *Plant Molecular Biology* (Shaw, C. H., ed.), IRL, Oxford and Washington DC, pp. 79–102.

4. Makaroff, C. A. and Palmer, J. D. (1987) Extensive mitochondrial specific transcription of the *Brassica campestris* mitochondrial genome. *Nucleic Acids Res.* **5,** 5141–5156.
5. Han, J. H., Stratowa, C., and Rutter, W. J. (1987) Isolation of full-length putative rat lysophospholipase cDNA using improved methods for mRNA isolation and cDNA cloning. *Biochemistry* **26,** 1617–1625.
6. Ye, F., Albaum, M., Markmann-Mulisch, U., and Abel, W. O. (1993) Improved method for the isolation of mitochondrial RNA from green leaves. *BioTechniques* **14,** 184.
7. Sambrook, J., Fritsch, E. F., and Maniatis, T. (1989) *Molecular Cloning: A Laboratory Manual,* 2nd ed. Cold Spring Harbor Laboratory Press, Cold Spring Harbor, NY.

8

Extraction of RNA from Fresh and Frozen Blood

Bimal D. M. Theophilus

1. Introduction

Whole blood contains nucleated white cells that constitute an easily accessible source from which RNA can be extracted, without the need for prior homogenization as is necessary with solid tissues. However, blood is a particularly problematic tissue from which to isolate RNA because RNA is extremely prone to degradation by ribonucleases, of which red cells are a rich source. Furthermore, blood constituents or their derivatives may inhibit PCR reactions *(1)*. RNA extraction from blood is therefore usually more successful if the nucleated white cells are first isolated from the red cells. As with extraction from other tissues, it is important to minimize degradation by following the appropriate recommendations for handling RNA, as detailed in the methodology below.

A variety of methods are employed for the extraction of RNA *(2)*; which usually comprise cell lysis, partitioning of RNA into a solvent fraction, and recovery of RNA from the solvent by precipitation.

The initial lysis is normally carried out in the presence of a protein denaturant, which simultaneously inactivates cytoplasmic nucleases that could degrade the RNA. Early methods employed phenol or phenol-chloroform to denature and precipitate proteins. RNA extracted using phenol may be contaminated with DNA or polysaccharides, which could interfere with subsequent manipulations, and may be removed by digesting with RNase-free DNase I or CsCl centrifugation.

The methods described below involve the acid-phenol-guanidinium method *(3)* and use a commercial reagent for fresh blood (RNAzol B; Biogenesis, UK) or a lab-prepared reagent (solution "D") for frozen archive whole blood. Both reagents contain guanidinium thiocyanate, a chaotropic agent that simultaneously disrupts cells and efficiently inactivates RNases.

A variety of kits and reagents are available from various companies for the isolation of RNA in addition to those described below, some of which are able to

From: *Methods in Molecular Biology, Vol. 86: RNA Isolation and Characterization Protocols*
Edited by: R. Rapley and D. L. Manning © Humana Press Inc., Totowa, NJ

isolate RNA, DNA, and protein from the same sample (*see* **Note 1**) (*see* Chapters 2–4, this volume). Since most research and diagnosis involves messenger RNA (mRNA), some methods have been developed only to isolate mRNA, either directly from the cell lysate or via an additional stage subsequent to the isolation of total RNA. This is achieved by exploiting the affinity of the poly(A) tail at the 3' end of the majority of mRNAs for a synthetic poly (T) sequence, which is usually immobilized on magnetic, cellulose, or silica beads *(4)* (*see* Chapter 13, this volume). However, most mRNA analyses including RT-PCR, can be performed starting with total cellular RNA, which is invariably easier to isolate and can be analyzed for quality on an agarose gel prior to subsequent analytical procedures.

RNA is extremely prone to degradation by contaminating ribonucleases. Degradation can be minimized by preparing aqueous solutions in double-distilled water containing 0.1% diethyl pyrocarbonate (DEPC; an inhibitor of RNases), which should then be incubated for at least 12 h at 37°C and sterilized. An exception to this is the Tris buffers, since DEPC reacts with amines. It is important that DEPC-treated solutions are autoclaved before contact with RNA since DEPC may chemically modify bases in RNA. Note: DEPC is toxic and should be handled in a fume cupboard.

Glassware and disposable plasticware such as pipet tips must be sterile, while nondisposable plasticware may be rinsed with chloroform or DEPC-treated water. In addition, it is recommended that a separate set of laboratory equipment is designated exclusively for RNA work.

RNase contamination from the hands of the investigator can be avoided by wearing and frequently changing dosposable gloves during manipulations.

2. Fresh Blood

2.1. Materials

1. X Phosphate-buffered saline (PBS).
2. Lymphoprep (Nycomed).
3. 10-mL centrifuge tubes (polypropylene or glass).
4. RNAzol B **(Note: RNAzol B contains guanidinium thiocyanate which is an irritant, and phenol which is a poison. It is therefore recommended that RNAzol B is handled in a fume cupboard.)**
5. Chloroform.
6. Isopropanol.
7. Ethanol.
8. 5*M* sodium chloride.

2.2. Method

1. Dilute 10 mL of anticoagulated whole blood 1:2 with 1X PBS in a sterile plastic 20 mL universal (*see* **Note 2**).

2. Carefully layer 10 mL of the diluted blood onto 3 mL of lymphoprep in each of 2 × 10 mL polypropylene or glass tubes able to withstand centrifugation at 800*g*. Ensure that a sharp interface is obtained with little or no mixing between the blood and separation fluid.
3. Centrifuge at 400*g* for 30–40 min or 800*g* for 15 min at room temperature (*see* **Note 3**).
4. Following centrifugation, a clear solution is obtained with aggregated erythrocytes sedimented to the bottom of the tube. Mononuclear cells, including lymphocytes form a distinct, cloudy band within the clear solution at the interphase of the upper sample plasma layer and the lower Lymphoprep solution (*see* **Fig. 1**). Transfer the mononuclear cell layer to a separate tube using a pipet tip or Pasteur pipet. The upper layer may first be removed to just above the band, if desired.
5. Make the cell solution up to 5 mL with 1X PBS, invert to mix and centrifuge as in **step 3**.
6. Decant the supernatant. The lymphocyte pellet may be stored for several days in this condition at –20°C before subsequent processing if desired.
7. Lyse the cells by the addition of 0.5 mL RNAzol B. Solubilize the RNA by passing the lysate through the pipet a few times.
8. Transfer the lysate to a sterile Eppendorf, add 50 μL of chloroform, shake samples vigorously for 15 s, and incubate on ice (or at 4°C) for 5 min. (Samples can be stored in this state for 1–2 h.)
9. Centrifuge the suspension at 12,000*g* for 15 min in a microfuge (*see* **Note 4**).
10. Transfer the upper aqueous phase (carefully avoiding the interphase, which contains DNA and proteins) to a fresh Eppendorf, add an equal volume of isopropanol (approx 400 μL), and store for 15 min at 4°C (or at –20°C overnight).
11. Microfuge samples at 12,000*g* for 15 min. The RNA pellet should be visible at the bottom of the tube.
12. Remove the supernatant and wash the RNA pellet once by adding 800 μL of 75% ethanol; vortex and centrifugation at 7500*g* for 8 min.
13. Resolubilize the RNA in 0.2*M* sodium chloride (e.g., 192 μL water + 8 μL 5*M* sodium chloride). Precipitate sample with 400 μL of 100% ethanol at –20°C for 1 h.
14. Centrifuge and ethanol wash as above (**steps 12** and **13**).
15. Allow the pellet to dry with tube open at room temperature for 15 min.
16. Solubilize the pellet in 50–100 μL of DEPC-treated water. The sample may be heated to 52–60°C for 5–15 min to aid solubilization.
17. Quantitate and analyze the RNA as described in **Subheading 3.3.**

3. Frozen Blood

3.1. Materials

1. 30-mL centrifuge tubes.
2. 1.5X solution D (6*M* guanidinium thiocyanate, 37.5 m*M* Na-citrate pH 7.0, 0.75% sarcosyl, 0.15*M* 2-mercaptoethanol).
3. Phenol-chloroform (5:1, pH approx 4.0).

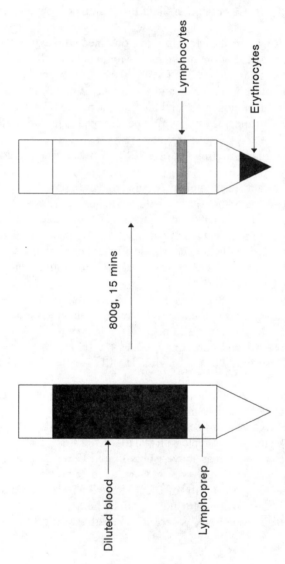

Fig. 1. Isolation of lymphocytes by centrifugation over Lymphoprep separation medium.

4. $2M$ sodium acetate, pH 4.0.
5. Isopropanol.
6. Ethanol.

3.2. Method

The method below is based on that described by Izraeli et al. *(5)*. The volumes given are based on a 2-mL frozen blood sample to be processed in a 30-mL centrifuge tube. Volumes in square brackets may be used to process a 5-mL sample without the need for a larger centrifuge tube.

1. Boil 2 mL (5 mL) of frozen whole blood in the collection tube in a water bath until it begins to thaw (5–25 s) (*see* **Note 5**).
2. As soon as the sample has thawed sufficiently to remove it from the tube (and before it has thawed completely) decant it into a sterile 30-mL centrifuge tube containing 3 mL (5 mL) of 1.5X solution "D" and vortex. Add 6 mL (10 mL) of phenol:chloroform (5:1) and 0.5 mL (1 mL) $2M$ sodium acetate pH 4; vortex after each addition. It is advisable to cover tubes with parafilm while vortexing.
3. Incubate on ice for 20 min.
4. Centrifuge at 5500g for 30 min at 4°C in a precooled rotor.
5. Transfer the supernatant to a separate centrifuge tube, or to several microfuge tubes, depending on the volume (*see* **Note 6**).
6. Add an equal volume of isopropanol and incubate the samples overnight at –20°C.
7. Centrifuge at 5500g for 30 min at 4°C.
8. Decant the supernatant and store the tubes inverted for 2–3 min.
9. Resuspend the pellet in 300 μL of 1.5X solution "D," transfer to an Eppendorf if necessary, add 400 isopropanol and incubate for 1 h at –20°C.
10. Centrifuge at 11,000g for 10 min at room temperature.
11. Decant the supernatant and wash the pellet by adding 750 μL of 80% ethanol followed by centrifugation at 11,000g for 2 min.
12. Remove the supernatant and resuspend the pellet in 50 μL DEPC-treated water.
13. Quantitate and analyze the RNA as described in **Subheading 3.3.**

3.3. Analysis of RNA

Determine the RNA quality and yield by reading the OD and OD_{260} and OD_{280} in a spectrophotometer. An OD_{260} of 1.0 corresponds to approx 40 g/mL of RNA; 1 mL of blood contains 5×10 nucleated cells, which usually yields about 1–3 μg of total RNA.

High quality RNA preparations that are free from contaminating DNA, protein, and polysaccharides should have an OD_{260}/OD_{280} ratio of 2.0 (*see* Chapter 12, this volume). Significantly lower values (<1.65) indicate contamination. Check the integrity of the RNA by running $^1/_{10}$ of the sample on a 1% agarose gel (**Fig. 2**) (*see* Chapter 11, this volume).

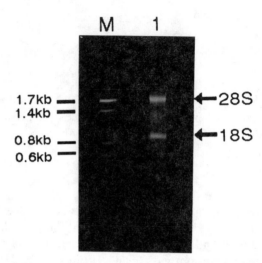

Fig. 2. Agarose gel (1%) of total cellular RNA extracted from fresh blood peripheral lymphocytes. M: double-stranded DNA molecular weight marker. Lane 1: Lymphocyte RNA showing 28S and 18S ribosomal RNA bands indicative of intact RNA. Depending on the extraction method employed, additional discrete small molecular weight bands (0.1–0.3 kb) may be visible corresponding to the small rRNA or tRNA populations.

The major cellular RNA species comprise 28S and 18S ribosomal RNAs (rRNA), which appear as defined bands if the RNA is intact. Degraded RNA appears as smaller molecular weight or absent rRNA bands with a large amount of low molecular weight smearing. Frozen blood generally yields RNA of poorer quality with variable degrees of background smearing apparent, although it may still be of adequate quality for RT-PCR. High molecular weight bands or smearing may be indicative of DNA contamination.

4. Notes

1. TRIzol reagent (Gibco-BRL Life Technologies, Edinburgh, UK) and RNA Isolator (Genosys, London, UK) are similar in composition and protocol to RNAzol B, but further enable DNA and proteins to be recovered by sequential precipitation following removal of the aqueous phase.

    The Hybaid Recovery RNA purification kit (Hybaid, London, UK) utilizes the acid-phenol-guanidinium method followed by adsorption to a binding matrix to isolate RNA. Hybaid also market Hybaid Recovery Amplification Reagent for the simultaneous isolation of RNA and DNA from whole blood.

    Nucleon HiPure and Nucleon QuickTrack (Scotlab, Palsley, UK) are pheonol-free extraction methods that enable recovery of RNA and DNA from the same sample and involve binding of proteinaceous material to a silica matrix.

The InVisorb TwinPrep (Bioline, UK) kit also enables the simultaneous isolation of RNA and DNA from the same sample. The procedure combines chaotropic agents with the binding of DNA to a mineral carrier. RNA is released into the supernatant from which it can be recovered by phenol/chloroform extraction followed by precipitation. DNA is eluted from the carrier in a low-salt buffer.

Catrimox-14 (VH Bio, UK; Iowa Biotechnology, Coralville, IA) provides a novel approach to RNA isolation by precipitating RNA complexed to a cationic surfactant as a reverse micelle that aggregates in water *(6)* (*see* Chapter 4).

United States Biochemical (Cleveland, OH) and Dynal (UK) market kits for the direct isolation of mRNA based on oligo(dT) cellulose and oligo (dT) magnetic beads, respectively.

Methods that simultaneously isolate DNA usually produce material of sufficient quality for PCR, but the DNA may not be of sufficient integrity to guarantee successful Southern blotting.

2. A suitable anticoagulant is 3.8% trisodium citrate diluted 1:10 into whole blood. Heparin may interfere with reverse transcription and PCR *(5)*, while EDTA interferes with cell lysis when using Catrimox-14.

3. g-Force can be converted into rpm using the formula $g = 0.0000118 \times r \times N^2$ where r = radius in cm and N = rpm.

4. Ideally, microfuge spins from this stage onwards should be carried out at 4°C, but may be performed at room temperature if a refrigerated microfuge is not available.

5. A glass tube should be used for efficient heat transfer when boiling the sample. Also, the tube should not narrow at the neck as this impedes removal of the partially thawed sample.

6. Occasionally, the interphase may be very diffuse with no aqueous phase visible. In this case, leave the samples at 4°C until the interphase has settled (overnight if necessary).

References

1. Higuchi, R. (1989) in *PCR Technology* (Erlich, H. A., ed.), Stockton, New York.
2. Jones, P., Qui, J., and Rickwood, D. (1994) *RNA Isolation and Analysis*. Bios Scientific, Oxford, UK.
3. Chomczynski, P. and Saachi, N. (1987) Single-step method of RNA isolation by acid guanidinium thiocyanate-phenol chloroform extraction. *Anal. Biochem.* **162,** 156–159.
4. Aviv, H. and Leder, P. (1972) Purification of biologically active globin messenger RNA by chromatography on oligothymidylic acid-cellulose. *Proc. Natl. Acad. Sci. USA* **69,** 1408–1412.
5. Izraeli, S., Pfleiderer, C., and Lion, T. (1991) Detection of gene expression by PCR amplification of RNA derived from frozen heparinized whole blood. *Nucleic Acids Res.* **19,** 6051.
6. Macfarlane, D. E. and Dahle, C. E. (1993) Isolating RNA from whole blood: The dawn of RNA-based diagnosis? *Nature* **362,** 186–188.

9

Isolation of Total RNA from Bacteria

John Heptinstall

1. Introduction

The susceptibility of RNA to degradation by exogenous and endogenous RNase activity following cell lysis has been well documented *(1,2)* . Moreover RNA usually occurs complexed with protein from which it must be released. Precautions to be taken against exogenous RNase include the use of plastic gloves, autoclaving solutions after adding 0.1% (v/v) diethyl pyrocarbonate (DEPC; except Tris, which reacts), and baking glassware, spatulas, and so forth at 180°C overnight *(3)*.

The endogenous RNase level varies with cell type, thus necessary precautions will vary. These may include the use of guanidinium thiocyanate (GuSCN), phenol, a thiol reagent (ß-mercaptoethanol, dithiothreitol), proteinase K, a detergent (sodium dodecyl [lauryl] sulphate, *N*-dodecyl sarkosine [sarkosyl]), placental RNase inhibitor (a protein found in placenta *[4]* and other tissues, and is sold under a variety of trade names), and vanadyl ribonucleoside complexes *(5)*. Some of the reagents (phenol, detergent, proteinase K, GuSCN) will also simultaneously deproteinize RNA.

In order to release nucleic acid from a bacterial cell, the membranes and peptidoglycan of the envelope must be disrupted. Commonly this is brought about by lysozyme/EDTA treatment *(6)*, with EDTA leading to a loss of lipopolysaccharide from the outer membrane of Gram-negative bacteria to allow access of lysozyme to peptidoglycan. The conditions are not always of sufficient duration to produce spheroplasts, which would completely lyse in the hypotonic solutions employed, unless genomic DNA is required. SDS is subsequently added, to inhibit RNase, to remove the cytoplasmic membrane, and also the protein that is complexed with RNA. GuSCN may be added at this stage, particularly if RNase activity is high *(7)*.

From: *Methods in Molecular Biology, Vol. 86: RNA Isolation and Characterization Protocols*
Edited by: R. Rapley and D. L. Manning © Humana Press Inc., Totowa, NJ

Following the addition of GuSCN, RNA may be separated from protein and DNA using phenol prior to precipitation with ethanol or isopropanol, although in extracting rRNA we have not found phenol treatment to be necessary. The pH must not be alkaline in view of the lability of RNA, and is normally 7.0 ± 0.2. Phenol treatment may be with buffer-saturated phenol alone, or with phenol-chloroform-isoamyl alcohol (25:24:1) in which subsequent phase separation is easier but both yield RNA in the upper, aqueous layer. Chloroform alone, without phenol, has been recommended for bacterial RNA extraction *(8)*. We have found this to give reasonably pure RNA, as judged by A_{260}/A_{280}, but very variable yields. The RNA is precipitated from solution by adding 0.1 vol $3M$ ammonium acetate, pH 5.2, and either 2.5 vol ethanol or 1.0 vol isopropanol at $-20°C$. Usually, following cell lysis, particularly if lysis is prolonged or vigorous, there will be contamination by DNA which will also be precipitated at this stage. Removal is effected with RNase-free DNase 1 (*see* **Note 1**) after which further phenol treatment may be necessary to remove the DNase.

If lysozyme was used for cell lysis it would be inactive in the presence of SDS, however proteinase K is not, even at $50°C$. Therefore a rapid method of disrupting bacterial cell envelopes, inactivating RNase, and freeing RNA, is to use proteinase K and SDS simultaneously *(9)*. This treatment may be followed by GuSCN, if residual RNase is a problem, or phenol treatment, or both. If phenol treatment is used, proteinase K will result in a reduction of the white protein precipitate at the aqueous/organic interface and higher RNA yields *(10)*. The one-step acid GuSCN phenol chloroform method for eukaryotic cells *(11)* uses GuSCN and mercaptoethanol and sarkosyl for cell lysis at pH 7.0, followed by sodium acetate at pH 4.0 (to $0.2M$) and water saturated (i.e., acidic) phenol. We have found with bacterial cells that cell disruption can be carried out directly by sonication in GuSCN solution. This is rapid but not applicable to lots of samples simultaneously and may lead to shearing of high-mol-mass RNA.

Polyethylene glycol is used to precipitate intact virus particles *(12)* and plasmid DNA *(13)* during plasmid or viral DNA preparation. It has been found that, following cell disruption, the cell debris and DNA can be removed by the addition of polyethylene glycol and salt, leaving the RNA in solution *(14)*. Subsequently, an aqueous biphasic system *(15)* may be produced by the addition of more salt (*see* **Note 2**), in which RNA selectively partitions into the lower, salt phase. Whether this is done or not, the RNA may then be precipitated with or without prior phenol treatment.

2. Materials

All solutions are prepared in water that has been previously autoclaved with 0.1% (v/v) DEPC. The latter is suspected of being a carcinogen and should be handled with care. Note: Care should be exercised when handling GuSCN and SDS, particularly in the solid state.

1. 5% (w/v) Sodium dodecyl sulphate (SDS).
2. 6M Guanidinium thiocyanate (GuSCN). Filter and store up to 1 mo at room temperature.
3. 4M Guanidinium thiocyanate, 25 mM sodium citrate, pH 7.0, 0.1M ß-mercapto–ethanol. Filter and store up to 1 mo at room temperature.
4. Proteinase K: 0.5 mg/mL. Aliquot and store at –20°C.
5. Physiological saline, 0.85% (w/v) NaCl. Autoclave.
6. Polyethylene glycol 6000 (PEG) 20% (w/v), 0.75% SDS (w/v) in 7.5% (w/v) potassium phosphate, pH 7.2. Autoclave.
7. Polyethylene glycol 6000 (PEG): 12.5% (w/v) in 1% (w/v) potassium phosphate, pH 7.2. Autoclave.
8. 0.12M Sodium phosphate buffer, pH 7.2. Autoclave.
9. Phenol, saturated with 0.12M phosphate buffer, pH 7.2. To phenol (either freshly redistilled at 165–180°C, or provided for use in molecular biology) in a previously unopened bottle, add phosphate buffer until the bottle is full. Mix gently. Add 0.05% (w/w of phenol) 8-hydroxyquinoline as an antioxidant and allow phase separation at 4°C. Remove and discard the upper, aqueous layer and repeat the buffer addition, mixing and decanting. Store the phenol at –20°C. Note: Phenol burns the skin. Should this occur, wash with 20% (w/v) PEG in 50% (v/v) industrial methylated spirits.
10. Chloroform (AnalaR grade).
11. 3M Sodium acetate, pH 5.2. Autoclave.
12. TBE buffer: Tris borate EDTA buffer. 90 mM Tris(hydroxymethyl)-aminomethane, 90 mM boric acid, 2.5 mM EDTA, pH 8.3. Prepare a stock 10X solution by dissolving 108 g Trizma base, 55 g boric acid, and 9.5 g disodium EDTA in water, to 1 L. Autoclave. Dilute 10X for use.
13. Agarose, 1.1% (w/v) in TBE. Autoclave. Agarose should be low or medium electroendosmosis (EEO).
14. TE buffer: 10 mM Tris-HCl, p H 8.0, 1 mM EDTA. Autoclave.
15. 10% (w/v) Sodium dodecyl sarcosinate (Sarkosyl). Autoclave.
16. L-broth. 10 g/L Tryptone, 5 g/L yeast extract, 5 g/L NaCl, 1 g/L glucose. Autoclave glucose separately.

3. Methods

3.1. Bacterial Culture

The protocol is written for *E. coli* B but is applicable to all *E. coli* strains tested. Amendments may be necessary for other bacteria (*see* **Note 3**).

1. Inoculate 50 mL of L-broth from a slope and grow overnight, 37°C.
2. Use 5 mL of this culture to inoculate 50 mL of L-broth and grow for 2.5 h at 37°C, to an absorbance (600 nm) of 0.45–0.60.

3. Centrifuge a 1.5 mL sample in an Eppendorf tube at 12,000g for 10 min. Discard the supernatant.
4. Wash in saline by resuspending the pellet in 300 µL of saline, centrifuge at 12,000g for 10 min and discard the supernatant.
5. Resuspend the pellet in 400 µL of 0.12 M sodium phosphate buffer, pH 7.2.

3.2. Lysis and Extraction with PEG 6000

1. To the resuspended pellet add 50 µL of 5% SDS and 50 µL of proteinase K (0.5 mg/mL). Vortex and incubate at 37°C for 20 min.
2. Add 500 µL of PEG 6000 (12.5% w/v) in 1% potassium phosphate, pH 7.2. Vortex and centrifuge at 12,000g for 10 min.
3. Carefully remove most of the supernatant, measuring the volume, into a sterile tube for precipitation of the RNA (*see* **Note 2**). If necessary, phenol extraction may be carried out at this stage (*see* **Subheading 3.2.7.**).
4. Add 0.1 vol 3M sodium acetate, pH 5.2, followed by 2.5 vol ethanol (or 1 vol isopropanol) at –20°C to precipitate the RNA. Leave for at least 1 h.
5. Centrifuge at 12,000g for 15 min, discard the supernatant, drain the tube by carefully inverting onto tissue paper. Wash the pellet (which may not be visible) with 70% ethanol by resuspending and centrifuging at 12,000g for 10 min. Decant and drain as before.
6. Dissolve the pellet in TE at room temperature for at least 30 min. The yield is about 15 µg total RNA/mL of culture (**Fig. 1**). The ratio A_{260}/A_{280} is in the range 1.90–2.05.
7. Should protein removal by phenol be required (**Subheading 3.2.3.**), add, sequentially, 0.1 vol 3M sodium acetate, pH 5.2 and 1 vol buffer-saturated phenol. Vortex and centrifuge at 12,000g for 10 min. Remove the upper, aqueous layer, repeat the extraction of this layer with 1 vol phenol, vortex, centrifuge and retain the upper layer. The RNA may be precipitated (**Subheading 3.2.4.**) without adding more sodium acetate (*see* **Note 4**).

We have not found endogenous RNase to be a problem with *E. coli* and some other organisms (*Pseudomonas aeruginosa, Klebsiella aerogenes*); however, should suppression of activity be required, then GuSCN should be incorporated into the extraction, either before or after cell lysis (*see* **Note 5**).

3.3. Lysis and Extraction with GuSCN

3.3.1. Addition After Cell Lysis

1. Cells are prepared and lysed, as in **Subheading 3.2.1.**
2. Add 750 µL of 6M GuSCN, vortex and centrifuge at 12,000g for 10 min. Transfer the supernatant to a fresh tube.
3. RNA can be precipitated directly at this stage, without significant protein contamination, by the addition of 0.1 vol sodium acetate and 2.5 vol cold ethanol, as **Subheading 3.2.4.** (**Fig. 1**).
4. Alternatively, phenol extraction may be carried out prior to precipitation (**Subheading 3.2.7.**).

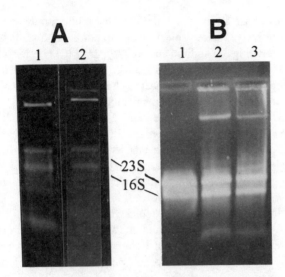

Fig. 1. Nondenaturing agarose gels of total RNA extracts from *E. coli* B. Precipitated with ethanol, but not phenol-treated. **(A)** 1.1% Agarose, about 2 µg nucleic acid per lane; lane 1, PEG; lane 2, GuSCN added *before* lysis. **(B)** 0.7% Agarose, about 6 µg nucleic acid per lane; lane 1, standard 23S & 16S rRNA; lane 2, PEG; lane 3, GuSCN added *after* lysis. A DNA band (not RNase-sensitive) is apparent between the 23S rRNA and the well in all cases shown here.

3.3.2. Addition Before Cell Lysis

Bacteria can be lysed by brief sonication in the presence of GuSCN. The procedure subsequently is then the one-step acid GuSCN-phenol method *(11)*, except that the detergent is added after sonication to avoid frothing.

1. Resuspend the bacterial pellet **(Subheading 3.1.4.)** in 1 mL of 4M GuSCN + sodium citrate + ß-mercaptoethanol and sonicate for 20 s.
2. Add 50 µL of 10% sarkosyl, vortex and centrifuge at 12,000g for 10 min.
3. The RNA may be precipitated at this stage by the addition of 0.1 vol 2M sodium acetate, pH 4.0 and either 2.5 vol ethanol or 1 vol isopropanol at –20°C followed by centrifugation at 12,000g for 10 min **(Fig. 1)**, or phenol-extracted in the presence of acidic sodium acetate and acidic phenol *(11)*.
4. Add, sequentially, 0.1 vol 2M sodium acetate, pH 4.0, 1 vol of water-saturated phenol and 0.2 vol chloroform isoamyl alcohol (49:1, v/v), vortex and centrifuge at 12,000g for 10 min.
5. The RNA in the upper, aqueous phase is precipitated by the addition of 2.5 vol ethanol or 1 vol isopropanol at –20°C (*see* **Subheading 3.2.4.**).

4. Notes

1. RNase-free DNase from some commercial suppliers may be less than perfect
   (16). Alternatively, various selective precipitation methods have been described
   for RNA, including 3*M* sodium acetate *(10)*, 2*M* LiCl *(10)*, or 0.5 vol ethanol *(7)*.
2. At this stage, with the RNA in PEG/salt solution, more PEG and salt can be
   added to produce an aqueous biphasic system *(15)*, with the RNA predominantly
   in the lower, salt phase. Thus, add an equal volume of 20% PEG 6000 + 0.75%
   SDS in 7.5% potassium phosphate, pH 7.2, mix for 30 s and centrifuge for 5 min
   at 12,000*g*. The upper, PEG-rich phase will occupy some 80% of the total volume
   and the salt-rich, lower phase 20%, but the latter contains about 80% of the total RNA.
3. Some organisms (*P. aeruginosa, K. aerogenes*) may be extracted under the
   conditions given, whereas others that have been tried (*Salmonella typhimurium,
   Proteus mirabilis, Serratia marcescens*) will require changes in some
   concentrations of reagents, in particular SDS and/or PEG may be increased.
4. If the RNA is not to be precipitated and washed, and should small amounts of
   phenol interfere with subsequent treatment of the RNA, then phenol can be
   removed by washing with chloroform. Add about an equal volume of chloroform,
   mix and centrifuge very briefly, discard the lower, chloroform layer. Repeat at
   least four times. Phenol may affect enzyme activity and certainly will give
   spuriously high A_{260}/A_{280} ratios since it gives a peak at 270 nm (*see* Chapter 12).
5. The efficacy of GuSCN as an inhibitor of RNase is dependent on the
   concentrations of both inhibitor and enzyme. Chaotropic effects are not
   apparent below 3*M* GuSCN, in some cases 5*M* may be required for sufficient
   inhibition *(17)*.

References

1. Barnard, E. A. (1964) The unfolding and refolding of ribonuclease in urea
   solutions 1. Rates and extents of physical changes. *J. Mol. Biol.* **10**, 235–262.
2. Aviv, H. and Leder, P. (1972) Purification of biologically active messenger RNA
   by chromatography on oligothymidylic acidcellulose. *Proc. Natl. Acad. Sci. USA*
   69, 1408–1412.
3. Sambrook, J., Fritsch, E. F., and Maniatis, T. (1989) *Molecular Cloning. A
   Laboratory Manual.* 2nd ed., Cold Spring Harbor Laboratory Press, Cold Spring
   Habor, NY, pp. 7.3–7.5.
4. Blackburn, P., Wilson, G., and Moore, S. (1977). Ribonuclease inhibitor from
   human placenta. Purification and properties. *J. Biol. Chem.* **252**, 5904–5910.
5. Berger, S. L. and Birkenmeier, C. S. (1979) Inhibition of intractable nucleases
   with ribonucleoside-vanadyl complexes: isolation of messenger ribonucleic acid
   from resting lymphocytes. *Biochemistry* **18**, 5143–5149.
6. Jones, P., Qiu, J., and Rickwood, D. (1994) RNA isolation and analysis. BIOS
   Scientific Publishers, Oxford, UK, pp. 15–28.
7. Chirgwin, J. M., Przybyla, A. E., MacDonald, R. J., and Rutter, W. J. (1979)
   Isolation of biologically active ribonucleic acid from sources enriched in
   ribonuclease. *Biochemistry* **18**, 5294–5299.

8. MacDonell, M. T., Hansen, J. N., and Ortiz-Conde, B. A. (1987) Isolation, purification and enzymatic sequencing of RNA. *Methods in Microbiology* **19**, 357–404.

9. Hilz, H., Wiegers, U., and Adamietz, P. (1975) Stimulation of proteinase K action by denaturing agents. Application to the isolation of nucleic acids and the degradation of masked proteins. *Eur. J. Biochem.* **56**, 103–108.

10. Lizardi, P. M. (1983) Methods for the preparation of messenger RNA. *Methods Enzymol.* **96**, 24–38.

11. Chomczynski, P. and Sacchi, N. (1987) Single-step method of RNA isolation by acid guanidinium thiocyanatephenolchloroform extraction. *Anal. Biochem.* **162**, 156–159.

12. Yamamoto, K. R., Alberts, B. M., Benzinger, R., Lawhorne, L., and Treiber, G. (1970) Rapid bacteriophage sedimentation in the presence of polyethylene glycol and its application to large-scale virus purification. *Virology* **40**, 734–744.

13. Sambrook, J., Fritsch, E. F., and Maniatis, T. (1989) *Molecular Cloning. A Laboratory Manual.* 2nd ed., Cold Spring Harbor Laboratory Press, Cold Spring Habor, NY, pp. 1.23–1.41.

14. Jenné, S., Miczka, G., and Heptinstall, J. (1993) Rapid extraction of bacterial ribosomal RNA with polyethylene glycol. Sixth European Congress on Biotechnology, Firenze, Italy, **3**, WE022.

15. Albertsson, P.-A. (1971) *Partition of Cell Particles and Macromolecules.* 2nd ed., Wiley-Interscience, New York.

16. Hengen, P. N. (1996) Methods and Reagents. *Trends Biochem. Sci.* **21**, 112,113.

17. Gillespie, D. H., Cuddy, K. K., Kolbe, T., and Marks, D. I. (1994) Dissolve and capture: a strategy for analysing mRNA in blood. *Nature* **367**, 390,391.

10

Isolation of Total RNA from Tissues or Cell Lines

Visualization in Gel

Tapas Mukhopadhyay and Jack A. Roth

1. Introduction

The purity and integrity of isolated RNA is a critical determinant of its effectiveness in such molecular biological procedures as Northern blot, poly A^+ RNA separation, cDNA synthesis, and in vitro transcription and translation. The successful isolation of total RNA from tissues or cell lines by any procedure involves four major steps: complete disruption of the cells or tissues; effective denaturation of the nuclear protein complex; inactivation of endogenous ribonuclease (RNase) activity; and purification of the RNA from the contaminating DNA, protein, and carbohydrates.

In mammalian cells, the majority of the RNA species (85%) are ribosomal RNA, comprising 28S, 18S, and 5S rRNA; 10% is made up of low-mol-weight RNA species (tRNA, small nuclear RNA, and so on) while only 1–3% is comprised of mRNA that encodes all of the polypeptides synthesized in the cell. The key steps in a good eukaryotic mRNA isolation procedure are to minimize RNase activity during the initial steps of the extraction process. Ribonucleases are highly stable and active enzymes that require no cofactor to function and are released immediately from cell membrane and membrane-bound organelles upon cell lysis. It is therefore essential to minimize RNase activity and to avoid introduction of any trace amount of RNase from all glassware, solutions, and handling.

Among the many methods used to extract total RNA from cultured cell lines or tissues, a combination of guanidinium thiocyanate and phenol chloroform extraction followed by isopropanol precipitation of the RNA is probably the

From: *Methods in Molecular Biology, Vol. 86: RNA Isolation and Characterization Protocols*
Edited by: R. Rapley and D. L. Manning © Humana Press Inc., Totowa, NJ

most widely used. Guanidinium thiocyanate is the most effective protein denaturant and has become the first choice for RNA extraction. The guanidinium method was successfully used by Chirgwin et al. *(1)* to isolate RNA from RNase-rich pancreas. The quality and quantity of the RNA isolated by this technique varies considerably depending on the cell and tissue types. In our laboratory, we essentially follow the acid guanidinium thiocyanate-phenol-chloroform method of Chomczynski *(2)*, with some modifications.

2. Materials

All materials should be RNase-free. RNases are ubiquitous and contamination from our hands and through coughing or other airborne dust particles can result in degradation of RNA. RNases are notoriously difficult to inactivate, even by boiling. To avoid such contamination, sterile microbiological techniques should be observed when handling solutions and reagents. Gloves should be worn at all times.

Sterile, disposable plasticware, when newly opened, is generally free from contaminating RNases. Glassware, except Corex tubes, should be baked at 200–300°C for at least 4 h. Corex tubes should be rendered RNase-free by treatment with diethyl pyrocarbonate (DEPC) for 1 h and autoclaved. All solutions should be prepared from chemicals which are molecular biology grade and, unless otherwise stated, treated with DEPC and autoclaved (*see* **Note 1**). Guanidinium thiocyanate is a potent chaotropic agent and irritant. Phenol is poisonous and causes severe burns. Proper laboratory clothing and handling of these reagents is required.

1. Guanidinine isothiocyanate solution. Dissolve 23:639 guanidinium thiocyanate to 25 mL DEPC treated water. Add 1.66 mL of 0.75*M* sodium citrate and filter sterilize using a 0.22 µm Nalgene filter assembly. Add 2.5 mL 10% lauryl sarcosine and 0.36 mL β-mercaptoethanol. Adjust to a final volume of 50 mL. This solution is kept at 4°C and is stable for 2 mo (*see* **Note 2**).
2. 2 *M* Sodium acetate, pH to 4.2 with acetic acid and filter sterilize.
3. Phenol:chloroform:ethanol at a ratio of 25:24:1.
4. Isopropanol.
5. RNase-free water: This can be achieved by treating the water first with diethyl pyrocarbonate (DEPC) at a concentration of 0.1%, incubating overnight at 37°C, followed by autoclaving to completely destroy any residual DEPC. DEPC is an efficient, nonspecific inhibitor of RNase; however, it is carcinogenic and therefore should be handled with extreme care.
6. 5*M* Lithium chloride.
7. 10% *n*-lauryl sarcosine I (made using DEPC-treated water).
8. 0.75*M* Sodium citrate, pH 7.0.
9. β-mercaptoethanol.

10. Vacuum grease, e.g., silicone lubricant (autoclaves).
11. Phosphate-buffered saline (available commercially) is used when starting material is cultured cells (*see* **Note 3**).
12. Phenol saturated with DEPC-treated water: This can be stored in darkness at 4°C for 1 mo.
13. Ethanol.
14. 70% Ethanol.
15. Chloroform-isoamyl alcohol (49:1).

3. Method

1. Suspend 1 g of tissue (*see* **Note 3**) or 10^9 cultured cells (*see* **Note 4**) in 4 mL guanidinium solution.
2. Add, sequentially, 0.4 mL of 2*M* sodium acetate (pH 4.0), 4 mL of water saturated phenol, and 0.8 mL of chloroform-isoamyl alcohol (49:1). Mix thoroughly by inversion.
3. Add 0.5 mL of silicone lubricant (*see* **Note 5**). Mix vigorously by vortexing for 1 min at room temperature.
4. Cool on ice for 30 min.
5. Transfer contents to a 50-mL thick polypropylene tube and centrifuge at 10,000*g* for 30 min at 4°C.
6. Remove the top aqueous phase (*see* **Note 5**) taking care to avoid disturbing the interface.
7. Add an equal volume of isopropanol to the aqueous phase. Mix and cool to –20°C. Incubate for at least 1 h (*see* **Note 6**).
8. Centrifuge at 10,000*g* for 20 min at 4°C to precipitate the RNA.
9. Remove the supernatant and solubilize RNA pellets in 500 µL of cold DEPC-treated water (*see* **Note 7**).
10. Add 2 mL of 5*M* LiCl solution to the RNA solution. Vortex for 10 s and incubate at 4°C for at least 2 h (*see* **Note 8**).
11. Precipitate RNA by low-speed centrifugation at 5000*g* for 20 min at 4°C.
12. Carefully remove the supernatant and solubilize the RNA pellet in 400 µL of guanidinium solution using an Eppendorf tube (*see* **Note 8**). Add 2 vol of absolute ethanol and precipitate RNA overnight at –20°C.
13. Precipitate RNA by centrifugation at 5000*g* for 10 min at 4°C.
14. Gently wash the RNA pellet in 70% cold ethanol and centrifuge at 5000*g* for 10 min at 4°C. Discard supernatant.
15. Dry the RNA pellet in a vacuum-dryer (*see* **Note 9**) and resuspend in DEPC-treated water or deionized formamide (*see* **Note 10**); *see* Chapter 12).
16. Check RNA purity by absorbance (*see* **Note 11**) and integrity by gel electrophoresis (*see* **Note 12** and Chapter 13).

4. Notes

1. Tris buffers cannot be treated with DEPC, since Tris reacts with DEPC and inactivates it.

2. Sometimes *n*-lauryl sarcosine may precipitate out of solution at 4°C; this can be resolubilized by heating to 65°C prior to use.

3. For tissue, snap freeze in liquid nitrogen and grind with a pestle (in the presence of liquid nitrogen) to form a fine powder. Add 4 mL of guanidinium solution. Transfer solubilized tissue to a Dounce glass Teflon homogenizer to ensure complete disruption of all material.

4. For cells grown in suspension, sediment cells (5×10^8 cells) in a 10 mL snap-cap polypropylene tube ($300g$ for 5 min at 4°C) and wash pellets in ice-cold phosphate-buffered saline. Reprecipitate cells ($300g$ for 5 min at 4°C), discard supernatant and solubilized cell pellets in 4 mL of guanidinium solution. For cells grown in a monolayer: Remove culture medium, wash with ice-cold sterile phosphate-buffered saline, and add 4 mL guanidinium solution directly to the culture flask. Solubilize cells with a scraper.

5. The addition of silicone lubricant helps to separate the upper aqueous phase, containing the RNA and the lower organic phase containing the DNA and protein. This makes it easy to collect the entire aqueous phase without contaminating the RNA with DNA or protein.

6. To avoid loss of RNA or to obtain the maximum RNA yield from samples containing relatively low amounts of RNA, an overnight precipitation is recommended.

7. The RNA at this stage is readily dissolved by pipeting.

8. RNA at this stage is often associated with carbohydrate moieties that can degrade RNA during storage and interfere with RNA quantitation (leading to unequal loading on agarose gels). The LiCl precipitation step eliminates such contaminants, helping to ensure RNA stability and purity.

9. Do not let the RNA pellet dry out, as this will make the pellet very diffcult to resuspend.

10. RNA is more soluble and stable in formamide and does not degrade with several freeze–thaw cycles. It can be stored (4 mg per 1 mL formamide) for at least 1 yr at −20°C and precipitated by adding 4 vol of ethanol. Moreover and in contrast with water solubilized samples, larger amounts of RNA solubilized in forma-mide can be applied to formaldehyde-agarose gels. Alternatively, RNA can be stored in DEPC-treated water at −80°C for periods greater than 1 yr.

11. The RNA prepared by this method is substantially free of contaminating DNA and protein and may be used directly for the Northern blot analysis, RNase protection assay, cDNA synthesis for reverse transcriptase-polymerase chain reaction, and translation in vitro. Purified RNA should show an $A_{260}:A_{280}$ ratio of 2 (*see* Chapter 12). However, because of variations in starting materials and indi-vidual practices the expected ratio ranges from 1.7 to 2. If the RNA preparation exhibits a ratio lower than 1.7, the RNA should be purified again. In most cases this is due to protein contamination and occurs when the aqueous phase is col-lected and some organic phase comes with it. The use of silicon grease, which forms a strong interphase between the organic and the aqueous phases, should minimize the mixing of the two phases and maximize the recovery of the aqueous phase. There are several methods of removing contaminating protein, the most

expedient is to re-extract the RNA with phenol-chloroform. Some loss of RNA may be expected from this additional step.

The isolated RNA should also exhibit an $A_{260}:A_{230}$ ratio greater than 2; *see* Chapter 12. A lower ratio indicates possible guanidinium thiocyanate contamination that can interfere with the enzymatic manipulation of RNA, e.g., cDNA synthesis or translation processes. This can be minimized by a second precipitation following the addition of 0.1 vol of $3M$ sodium acetate (pH 5.3) and 2 vol of absolute alcohol, mixing and cooling to $-20°C$ overnight. The mixture is centrifuged at $5000g$ for 15 min at $4°C$ and washed with 70% cold ethanol. It is then dried and resuspended in DEPC-treated water. (If the RNA concentration is very low, then centrifugation should be done at $10,000g$ for 30 min, but the resulting compacted RNA pellet may be difficult to dissolve.)

12. The integrity of the purified RNA is determined by denaturing agarose gel electrophoresis. Several methods are suitable for this purpose. The use of formaldehyde gel electrophoresis has been described (*see* Chapter 13).

Smearing of RNA is an indication of degradation. Although some tissues contain high levels of endogenous RNases, this RNA extraction protocol described above works well for most cell lines and tissue samples. Ensure that all materials used are RNase-free.

If equal amounts of RNA (as determined by O.D. measurements) appear unequal when visualized by UV and ethidium bromide staining, carbohydrate contamination is the most likely reason. This can be minimized by repeating the lithium chloride precipitation step. This significantly increases the stability of the RNA by removing contaminating materials that can trap RNase. This problem is more common in mammary gland tissues and some cell lines. Washing the RNA pellet with cold 70% ethanol can also clean up residual contaminants.

References

1. Chirgwin, J. M., Przybyla, A. E., MacDonald, R. J., and Rutter, W. J. (1979) Isolation of biologically active ribonucleic acid from sources enriched in ribonuclease. *Biochem. J.* **18**, 5294.
2. Chomczynski, P. and Sacchi, N. (1987) Methods of RNA isolation by acid guanidinium thiocyanate-phenol-chloroform extraction. *Anal. Biochem.* **162**, 156–159.

11

Isolation of Messenger RNA

Sian Bryant and David L. Manning

1. Introduction

mRNA comprises approx 1–5% of total cellular RNA. Although the actual amount depends on the type of cell and its physiological state, at any given time approx 12,000 genes are being transcribed with approx 500,000 mRNA molecules present in each mammalian cell.

Eukaryotic mRNAs are heterogeneous in size (ranging from 0.5 kb to over 20 kb) and abundance (from fewer than 15 copies to over 20,000 copies per cell). The presence of a terminal stretch of approx 200 adenosine residues (the polyA tail) on most eukaryotic mRNAs and its absence in ribosomal and transfer RNAs has important practical consequences, since it allows polyadenylated species (messenger RNAs) to be separated from their nonpolyadenylated counterparts (ribosomal and transfer RNAs that account for over 90% of total cellular RNA).

High-quality mRNA is needed for a number of molecular biology techniques, including cDNA library construction (1,2). Not surprisingly, numerous mRNA extraction kits are now commercially available. All exploit the same basic principle, described in this chapter, which involves the affinity selection of polyadenylated mRNA using oligodeoxthymidylate (oligo [dT]).

2. Materials

All materials used in this procedure should be sterile and of molecular biology-grade. All Tris-containing solutions are prepared using RNase-free water and autoclaved. All other solutions, unless otherwise stated, should be treated directly with diethyl pyrocarbonate (DEPC) and autoclaved. DEPC is an efficient, nonspecific inhibitor of RNase activity. It is, however, a carcinogen and should be handled in a fume hood with extreme care. The hands are a major source of RNase activity and gloves should be worn for all procedures.

From: *Methods in Molecular Biology, Vol. 86: RNA Isolation and Characterization Protocols*
Edited by: R. Rapley and D. L. Manning © Humana Press Inc., Totowa, NJ

1. RNase-free water: Add 0.1% (DEPC) to water. Allow to stand overnight at 37°C and autoclave to destroy residual DEPC activity. All solutions except Tris, which inactivates DEPC, can be treated in the same way.

2. SDS (sodium dodecyl sulphate): SDS is dangerous by inhalation and should be weighed in a fume hood. A 10% stock solution is normally prepared. This solution is unstable if autoclaved, however any residual RNase activity can be destroyed by heating the solution at 65°C for 2 h.

3. Oligodeoxthymidylate-cellulose (oligo[dT]): Oligo (dT) cellulose is available commercially. Although the binding capacity of oligo(dT) cellulose varies between different suppliers, a general rule is to use 25 mg of oligo(dT) for each 1 mg of total RNA.
   Suspend oligo (dT) cellulose in loading buffer at a concentration of 5 mg/mL loading buffer. Oligo (dT) is insoluble and should be resuspended by gentle tapping or inversion. Do not vortex. It can be stored either dry at 4°C or suspended in loading buffer at −20°C.

4. RNase-free glass wool and Pasteur pipets: Wrap both the glass wool and Pasteur pipets in aluminium foil and bake at 200°C for 2–4 h to remove any RNase activity.

5. 5M NaCl: Store at room temperature.

6. 3M Sodium acetate pH 6.0: Store at room temperature.

7. Absolute alcohol. Store at −20°C.

8. 70% Ethanol. Prepare this solution using DEPC-treated water. Store at 4°C.

9. Loading buffer: 0.5M NaCl in 0.5% SDS, 1 mM EDTA, 10mM Tris-HCl, pH 7.5 (*see* **Note 1**). Store at room temperature.

10. Elution buffer: 1 mM EDTA, 10mM Tris-HCl, pH 7.5. This can be stored at room temperature but should be preheated to 65°C prior to use.

11. Recycling buffer: 0.1M NaOH. This should be prepared immediately before use and used fresh.

3. Methods
3.1. Preparing an Oligo (dT) Column

Oligo (dT) columns are available commercially or can be prepared by using a 1–3 mL syringe. Preparing your own columns is both easy and cheap.

Remove the plunger from the syringe and plug the base with glass wool. Add Oligo (dT) cellulose to the syringe using a sterile RNase-free Pasteur pipet. The oligo (dT) cellulose will collect, as a column, above the glass wool. The loading buffer will escape through the glass wool and can be discarded. To ensure the oligo (dT) cellulose is packed and free from air locks, add 3 vol of loading buffer using a pipet and allow the solution to run through the column. The column is now ready for immediate use and should not be run dry.

3.2. Isolation of Poly(A+) RNA

1. Resuspend the RNA pellet in loading buffer or if in solution, add 1/10 vol of 5M NaCl (*see* **Note 1**).

2. Heat-denature RNA and immediately load onto the column (*see* **Note 2**). Allow the RNA solution to enter the column and then apply 3 vol of loading buffer.
3. Re-apply the eluate to the column (*see* **Note 3**).
4. Wash with 3 vol of loading buffer (*see* **Note 4**). Discard eluate.
5. Recover the bound poly(A$^+$) mRNA by adding 3 vol elution buffer. Collect the mRNA in a sterile tube on ice (*see* **Note 5**).
6. mRNA is precipitated by adding 1/10 vol of 3M sodium acetate and 2 vol of ice-cold absolute ethanol. An overnight precipitation at –20°C maximizes the precipitation of RNA.
7. Centrifuge at 15,000g for 15 min to pellet the RNA. Discard the supernatant.
8. Wash the RNA pellet in ice-cold 70% ethanol (*see* **Note 6**). Centrifuge at 15,000g for 5 min to repellet the RNA, which may have been disturbed by washing. Discard the supernatant.
9. Dry the RNA pellet. Once dry, resuspend in DEPC-treated water.
10. Assess the purity and integrity of mRNA (*see* **Note 7**).

4. Notes

1. SDS can be omitted from the loading buffer as it may precipitate in cold or air-conditioned laboratories and clog the column. Residual SDS may coprecipitate with the RNA and interfere with other procedures, such as reverse transcription.
2. RNA can be linearized or denatured by heating to 80–90°C for 5 min, cooling quickly on ice (taking care not to cause precipitation of loading buffer components) and applying the solution to the column immediately. This reduces RNA secondary structure and aids binding of the poly (A$^+$) tailed mRNA to the oligo (dT) column.
3. This step increases the yield of poly(A$^+$) RNA.
4. The poly A tail anneals to Oligo (dT) in the presence of high salt (NaCl) concentrations. The further addition of 3 vol of loading buffer ensures nonpolyadenylated RNA species are washed from the column.
5. The oligo (dT) column can be regenerated by washing with 10 vol of recycling buffer followed by re-equilibration with three volumes of loading buffer. Oligo (dT) can be stored dry at 4°C or resuspended in loading buffer at –20°C until required.
6. This step removes any contaminating salt that may have coprecipitated with the mRNA.
7. mRNA purity is measured by absorbance at 260 nm. (An absorbance of 1.0 at 260 nm is equivalent to 40 mg mRNA). An additional reading at 280 nm allows the A_{260}:A_{280} ratio to be calculated. A ratio of 2.0 should be expected using this protocol. The integrity of the mRNA can be assessed by gel electrophoresis as described in Chapter 13. mRNA should appear on ethidium bromide stained gels as a smear ranging from 200 bp to greater than 10 kb with no detectable ribosomal RNA. If small amounts of mRNA are added to the gel, visualization by ethidium bromide may be impossible. To circumvent this problem, set up a Northern blot (as described in Chapter 17) and hybridize using a labeled oligo(dT) primer. Autoradiography should reveal mRNA fragments ranging in size as described above.

References

1. Manning, D. L., Daly, R. J., Lord, P. G., Kelly, K. F., and Green, C. D. (1988) Effects of Estrogen on the expression of a 4.4 kb mRNA in the ZR-75-1 human breast cancer cell line *Mol. Cell. Endo.* **59** 205–212.
2. Manning, D. L., Archibald, L. H., and Ow, K. T. (1990) Cloning of estrogen-responsive mRNAs in the T-47D human breast cancer cell line. *Cancer Res.* **50** 4098–4104.

12

UV Spectrophotometric Analysis of Ribonucleic Acids

Ralph Rapley and John Heptinstall

1. Introduction

The ability to quantify nucleic acids accurately and rapidly is a prerequisite for many of the methods used in biochemistry and molecular biology. In the majority of situations this is carried out using spectrophotometry, which is nondestructive and allows the sample to be recovered for further analysis or manipulation. Spectrophotometry makes use of the fact that there is a relationship between the absorption of ultraviolet light by RNA and its concentration in a sample. The absorption maximum of RNA is approx 260 nm. This figure is an average of the absorption of the individual ribonucleotides that vary between 256 nm and 281 nm. In the case of RNA the concentration of a sample containing RNA may be calculated from the following equation:

$$40 \times OD_{260} \text{ of the sample} = \text{concentration of RNA (µg/mL)}$$

That is when the OD_{260} of the sample is 1 the concentration of RNA will be approx 40 µg/ml. It is also possible to assess the degree of purity of the RNA by examining the absorption at other wavelengths in which protein and polysaccharides have known absorption maxima. Proteins are known to absorb strongly at 280 nm and polysaccharides may be identified by their maximum at 230 nm. Therefore the ratio of measurements of these three wavelengths 230, 260, and 280 may indicate the degree of purity of the RNA. For a sample containing only RNA this is judged as being the ratio 1:2:1. If there is significant deviation from this then it is evident that contaminants are present and that further purification of the sample is necessary. In many cases the purity and the concentration may be further obscured by the presence of reagents that are used in the extraction process itself. Some of these have characteristics that

From: *Methods in Molecular Biology, Vol. 86: RNA Isolation and Characterization Protocols*
Edited by: R. Rapley and D. L. Manning © Humana Press Inc., Totowa, NJ

are evident on a spectrophometric scan that includes the three wavelengths indicated. Therefore when using spectrophotometry in the analysis of RNA it is necessary to be aware of the potential problems that may result in misleading figures *(1)*.

2. Materials (*see* Notes 1 and 2)

1. Spectrophotometer capable of scanning in the ultraviolet wavelength range.
2. Optically matched quartz cuvettes.
3. 0.1% diethyl-pyrocarbonate.
4. Molecular biology grade H_2O (ribonuclease-free).

3. Methods (*see* Notes 3 and 4)

3.1. Performing Spectrophotometry

1. Prepare the cuvettes by soaking them in 0.1% diethylpyrocarbonate for at least 15 min.
2. Carry out baseline correction with H_2O or a non-UV-absorbing buffer.
3. Dilute the RNA sample with H_2O or a non-UV-absorbing buffer and carry out analysis.
4. Record absorption for 230 nm, 260 nm, and 280 nm and obtain a scan of absorption between 200 nm and 300 nm.

3.2. Analysis of Spectrophotometric Scans

Figures 1–4 indicate typical scans of samples containing RNA in addition to some of the reagents commonly used to extract them *(2)*. It is evident that in some cases it is necessary to scan samples to provide an indication of the degree of contamination rather than merely taking fixed readings at 260 nm and 280 nm. **Figure 1** indicates a trace of 33 µg/mL of yeast tRNA (**Fig. 1A**) overlaid by a second trace (**Fig. 1B**) in which the sample also contains 16 µL/mL of phenol. It is clear that trace amounts of phenol remaining after an extraction can affect the $OD_{260:280}$ ratio and the subsequent calculation to obtain the concentration of the RNA. In this case, the $OD_{260:280}$ for **Fig. 1A** is 2.1:1, whereas for **Fig. 1B** the ratio is 2. This phenomenon is exaggerated if the RNA concentration is lower and/or the phenol concentration is increased as evident in **Fig. 2** where **Fig. 2A** is the pure RNA scan (16 µg/mL) and **Fig. 2B** has an additional 32 µL/mL of phenol. In this case, as with Fig. 1, the $OD_{260:280}$ for **Fig. 2B** is misleading (1.87:1), indicating acceptably pure RNA.

Figure 3A is a trace of 16 µg/mL of RNA, **Fig. 3B** contains an additional 132 µM guanidinium thiocyanate. There appears to be little difference in scans and the $OD_{260:280}$ ratios are identical (2.1:1). However, increasing the guanidinium thiocyanate concentration from millimolar to molar amounts as used in a typical extraction and indicated in **Fig. 3C** has a drastic effect. **Figure 4A** indicates a scan of 16 µg/mL of pure RNA, but **Fig. 4B** contains 6.25% (w/v) polyethyleneglycol. This has the effect of masking the true concentration of RNA in the sample,

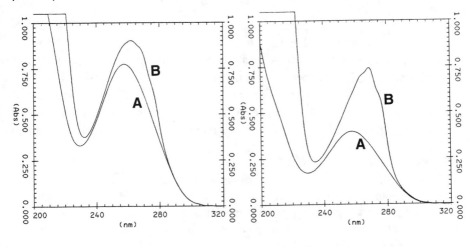

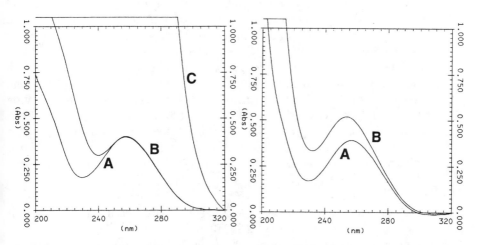

Fig. 1. *(top left)* Results of a spectrophotometric scan from 200–320 nm of a solution containing: **(A)** 33 µg/mL of tRNA alone (OD_{260}:OD_{280}) 2.1:1; **(B)** also contains the same concentration of tRNA, however 16 µL/mL of phenol is present representing a 1.6×10^{-3}% solution, (OD_{260}:OD_{280}) 2.03:1.

Fig. 2. *(top right)* Results of a spectrophotometric scan from 200–320 nm of a solution containing: **(A)** 16 µg/mL of tRNA (OD_{260}:OD_{280}) 2.16:1; **(B)** contains the same concentration of RNA, however, 32 µL/mL of phenol is present, (OD_{260}:OD_{280}) 1.87:1.

Fig. 3. *(bottom left)* Results of a spectrophotometric scan from 200–320 nm of a solution containing: **(A)** 16 µg/mL tRNA (OD_{260}:OD_{280}) 2.3:1; **(B)** contains the same concentration of RNA, however, 132 µM of guanidinium thiocyanate is present (OD_{260}:OD_{280}) 2.1:1; **(C)** contains 2.4 *M* guanidinium thiocyanate and the (OD_{260}:OD_{280}) cannot be measured.

Fig. 4. *(bottom right)* Results of a spectrophotometric scan from 200–320 nm of a solution containing: **(A)** 16 µg/mL tRNA (OD_{260}:OD_{280}) 2.3:1; **(B)** contains the same concentration of RNA, however 6.25% (w/v) polyethylene glycol is present, (OD_{260}:OD_{280}) 2.5:1.

since the $OD_{260:280}$ ratios are similar at 2.3 and 2.5 respectively, but the RNA concentration would be calculated on an OD_{260} of 0.38 (**Fig. 4A**) or 0.48 (**Fig. 4B**). From this information it is clear to see that when analyzing ratios and concentrations of RNA spectrophotometrically it is necessary not only to derive readings at 280, 260, and 230 nm but also to scan throughout the range 200 nm to 300 nm. Trace amounts of reagents used in the extraction process can influence adversely and provide misleading data that may affect any subsequent manipulation.

4. Notes

1. It is advisable to refer to manufacturer's instruction for the particular type of spectrophotometer used, a number of newer machines are linked to PC computers and give readings at the required wavelengths.
2. Gloves should be worn throughout the procedure to minimize the effect of any exogenous RNase activity.
3. It is always advisable when extracting and analyzing RNA to perform gel electrophoresis in addition to spectrophotometry. This gives an indication of the integrity of an extraction since in many cases it is possible to visualize a major RNA species such as 16/23S rRNA for prokaryotes and 18/28S for eukaryotes.
4. It is also possible to determine approximate RNA concentrations using spot assessments with ethidium bromide. Cover a transilluminator with plastic film such as Saran Wrap. Pipet 2–5 μL of RNA samples of known concentration (0–30 μg/mL) directly onto Saran Wrap. Next to these add the unknown RNA sample. To all samples add ethidium bromide (2 μg/mL in standard TE buffer (10 m*M* Tris-HCl, 1 m*M* EDTA [pH 7.6]). It is possible to visualize the RNA and make an estimation of the RNA or alternatively record the results on a polaroid camera.

References

1. Manchester, K. L. (1995) Value of A_{260}/A_{280} ratios for measurement of purity of nucleic acids. *BioTechniques* **19**, 208–210.
2. Heptinstall, J. (1997) Isolation of bacterial RNA, in *Methods in Molecular Biology, Vol. 86:* RNA Isolation and Characterization Protocols (Rapley, R. and Manning, D. L., eds.), Humana, Totowa, NJ.

13

Formaldehyde Gel Electrophoresis of Total RNA

Sian Bryant and David L. Manning

1. Introduction

RNA has the tendency to form both secondary and tertiary structures that can impede its separation by electrophoresis. As such, identical species of RNA exhibiting varying degrees of intramolecular base-pairing, migrate at different rates and result in the smearing of distinct RNA molecules. Consequently, the electrophoresis of RNA needs to be performed under denaturing conditions. Heat denaturing the RNA sample prior to electrophoresis is insufficient, as secondary structures will simply reform unless a denaturing system is used.

Successful electrophoresis of RNA is therefore accomplished in two steps: RNA is heat denatured prior to electrophoresis; and during electrophoresis, conditions are established that maintain the RNA in a denatured state.

The methodology described in this chapter involves the use of formaldehyde as a denaturant within the agarose gel. In addition, both formaldehyde and formamide are added to the sample before electrophoresis to aid the denaturation of the RNA sample. For procedures such as Northern analysis, in which RNA is transferred or blotted from the gel to a solid matrix for subsequent hybridization (*see* Chapters 15 and 17), the optimal balance between electrophoretic resolution and efficiency of transfer is achieved with a 1–1.2% agarose gel (*see* **refs.** *1–3*).

2. Materials

It is very important that all reagents used are of molecular biology-grade and free from RNase contamination. As in all molecular biological procedures, gloves should be worn throughout.

1. Formaldehyde: Formaldehyde is a suspected nose, nasopharynx, and liver carcinogen; it is toxic both through inhalation and ingestion. Its use should

From: *Methods in Molecular Biology, Vol. 86: RNA Isolation and Characterization Protocols*
Edited by: R. Rapley and D. L. Manning © Humana Press Inc., Totowa, NJ

therefore be restricted to a ventilated fumehood. Formaldehyde is routinely supplied as a 37% (v/v) stock solution. It should be stored at room temperature and out of direct sunlight to prevent oxidation.

2. Formamide: The use of this chemical should be restricted to the fumehood, as it is a suspected teratogen and is irritating to the eyes, skin, and respiratory system. Formamide should be stored in the dark and at room temperature to prevent oxidation.

3. RNase-free water: This can be achieved by treating the water first with diethyl pyrocarbonate (DEPC). Add DEPC to water to a final concentration of 0.1%. Incubate overnight at 37°C and autoclave to destroy any residual DEPC. DEPC is an efficient, nonspecific inhibitor of RNase, however it is carcinogenic and should be handled in a fume hood with extreme care.

4. 10X MOPS buffer: (0.2M MOPS (3-(N-Morpholino)propanesulfonic acid) pH 7.0; 50 mM sodium acetate, pH 6.0; 10 mM EDTA, pH 8.0. This solution is prepared using RNase-free water. When autoclaved it assumes a characteristic golden color and can be stored at room temperature.

5. Agarose (molecular biology–grade): Store at room temperature.

6. Ethidium bromide solution (10 mg/mL): This is a powerful mutagen, extreme care must be taken when using this substance. Store at 4°C.

7. Horizontal electrophoresis tank and a 11 × 14 cm casting tray.

8. RNA-loading buffer: 50% glycerol, 1 mM EDTA and 0.4% bromophenol blue, made up to the required volume with DEPC-treated water.

3. Method

1. Denature the RNA in a sterile RNase-free microcentrifuge tube by mixing the following:

   10 μg of RNA (in a final volume of 5 μL with DEPC-treated water),
   2 μL of 10X MOPS solution,
   3.5 μL of formaldehyde,
   10 μL of formamide.

   Incubate the RNA solution at 65°C for 15 min in a fumehood and transfer immediately to ice.

2. For a 1% agarose gel using an 11 × 14 cm casting tray, mix 0.8 g of agarose with 57.5 mL of sterile RNase-free water and boil to dissolve agarose. Cool to 60°C, then add 8 mL of 10X MOPS buffer, 14.5 mL of 37% (12.3M) formaldehyde and finally 8 μL of ethidium bromide solution (at a concentration of 10 mg/mL). Mix the components of the gel by gently rotating the bottle, being careful not to introduce any air bubbles. Pour the gel into the prepared casting tray (in a fume hood), insert the comb and let it set for a least 60 min at room temperature.

3. After cooling the denatured RNA solution in ice, add 2 μL of sterile RNA loading buffer.

4. Fill the buffer reservoirs and cover the gel with 1X MOPS buffer.

5. Pre-electrophorese the gel at 50 V for 10 min.

6. Load the heat denatured samples into the wells and run the gel at 50 V until the

bromophenol blue has moved approximately half to three quarters of the way along the gel (*see* **Note 1**).

7. Visualize RNA using a 254 nm short wave ultra-violet transilluminator (*see* **Notes 2–7**). Gels can be photographed using a polaroid land camera with 66S Polaroid film.

4. Notes

1. To avoid overheating and the generation of a smile effect on the migration of RNA and bromophenol blue dye, we routinely run gels at 4 V per cm length of gel (i.e., for a 14-cm long gel, electophoresis is performed at 50–60 V for approx 4 h).
2. RNA bands very broad or trailing. This problem usually occurs as a result of a fault in the sample preparation (RNA isolation) or in the loading of the gel. The salt composition (introduced with the sample) of the loading solution may be too high, too much RNA may have been loaded into the well (slight band broadening may occur with volumes above 30 μL or more than 50 μg of RNA per well); and applying higher voltages when running the gel may cause broadening and trailing probably due to excess heat. Electrophoresis may be carried out quite successfully at room temperature.
3. Little or no RNA detected in the gel after electrophoresis. RNA tends to aggregate at the top of the gel, close to the well, if it is contaminated with proteins. Normally, however, this is prevented since most procedures for RNA extraction and purification involve the use of proteolytic enzymes or deproteinizing mixtures such as phenol:chloroform. RNA may be degraded into fragments that are so small they pass straight through the gel, if this is the case run the RNA on a gel containing a higher concentration of agarose or preferably run the gel for a shorter period of time. If there is a problem with RNA degradation, then the extraction and purification methods need to be examined. Degradation of the RNA could be a result of ribonucleases in the electrophoresis buffer or tank, therefore care must be taken to ensure that the equipment is clean and both this and the buffer are nuclease-free. The gel concentration may be too high, preventing the RNA from entering it, although this is very unlikely as the concentration used in this methodology is not very high.
4. There may be problems with ethidium bromide-staining in formaldehyde gels. It can retard the running of the nucleic acid on the gel by up to 15%, but more importantly it can reduce the transfer efficiency of the RNA to a solid support. Ethidium bromide can be removed by soaking the gel in running buffer or transfer buffer, with several changes, for 1 h.
5. An aliquot of total cellular RNA should electrophoretically resolve only two very distinct bands, the 28s and 18s rRNAs. The appearance of these two bands gives an idea of the integrity of the RNA. The mRNA component manifests itself as a significantly lighter stained smear above, between and below the rRNA bands. The 5s rRNA, 5.8s rRNA, and the tRNAs all migrate close to the leading edge of the gel. Totally degraded samples appear as heavily localized smears below the level at which the 18s rRNA appears in intact samples. Heavy smears that appear along the length of the lane may be indicative of degradation or may simply mean

the sample was incompletely denatured. Fluorescence within the well suggests that genomic DNA is present within the sample.

6. The positioning of the 18s and 28s bands should be noted as this information may be required to estimate the size of a particular RNA of interest (18s rRNAs range in size from 1.8 to 2 Kb, 28s rRNAs range in size from 4.6 to 5.2 Kb).

7. Prolonged exposure to UV light can damage the RNA. Formaldehyde can be removed from the gel by placing in either 1X MOPS buffer or DEPC-treated water for about 15 min, with several changes of buffer.

References

1. Boedtker, H. (1971) Conformation-independent molecular determinations of RNA by gel electrophoresis. *Biochim. Biophys. Acta.* **240**, 448.

2. Rave, N., Crkvenjakov, R., and Boedtker, H. (1979) Identification of procollagen mRNAs transferred to diazobenzyloxymethyl paper from formaldehyde gels. *Nucleic Acids Res.* **6**, 3559.

3. Manning, D. L., McClelland, R. A., Gee, J. M., Chan, C. M., Green, C. D., Blamey, R. W., and Nicholson, R. I. (1993) The role of four estrogen responsive genes pLIV1, pS2, PSYD3, and pSYD8 in predicting responsiveness to therapy in primary breast cancer. *Eur. J. Cancer* **29(10)**, 1462–1468.

14

Preparation of RNA Dot-Blots

Rachel Hodge

1. Introduction

RNA dot hybridizations were first described by Kafatos et al. *(1)*. They enable rapid detection of transcription from a number of different mRNA populations and are particularly useful in the initial characterization of cDNA clones isolated by differential screening. In cases in which many samples have to be handled, filtration manifold systems are available, such as the Millipore MilliBlot system, which use a vacuum source to transfer nucleic acid to filter.

Although it is possible to use a pure nitro-cellulose membrane matrix, the nylon-based nitro-cellulose coated membranes currently available are much easier to handle. In addition, nitro-cellulose membranes are unsuitable for use with the digoxigenin/AMPPD system.

This protocol is a slightly modified version of that supplied with Hybond N (Amersham International, Amersham, UK). RNA is first diluted in a solution containing a MOPS buffered formamide/formaldehyde mixture. The solution is then heated to remove secondary structure and cooled rapidly on ice. After addition of salt, the RNA solution is dotted onto the membrane and fixed to it by UV crosslinking.

2. Materials

All chemicals used should be AnalaR–grade. All aqueous solutions should be DEPC (diethyl pyrocarbonate)-treated prior to autoclaving (*see* **Note 1**), to destroy ribonuclease activity. Gloves should be worn at all times particularly when handling membranes and RNA solutions.

1. Nylon membrane such as Hybond N (Amersham International).
2. DEPC-treated water.
3. 10X SSC: 1.5M Sodium chloride, 0.15M tri-sodium citrate.

From: *Methods in Molecular Biology, Vol. 86: RNA Isolation and Characterization Protocols*
Edited by: R. Rapley and D. L. Manning © Humana Press Inc., Totowa, NJ

4. RNA: Total RNA or polyA$^+$ RNA can be used. RNA should be checked for degradation by gel electrophoresis. A concentration of 10 mg/mL or above is ideal (*see* **Note 2**).

5. 10X MOPS: 0.2M 3-[N-morpholino] propanesulphonic acid sodium salt, 90 mM sodium acetate, 10 mM EDTA di-sodium salt, to pH 7.0 with sodium hydroxide. Store in the dark at 4°C.

6. RNA incubation solution: To prepare 1 mL mix: 657 µL formamide, 210 µL 37% formaldehyde solution (37% w/v as supplied) and 133 µL 10X MOPS (*see* **Note 3**). This solution may be prepared fresh or stored at –20°C indefinitely. Both formamide and formaldehyde are toxic and should be handled in a fumehood.

7. Hair dryer (optional).

3. Method

1. Cut the membrane to a suitable size (allow 1 cm^2 per RNA dot) and mark (with a pencil) to show orientation and positions for sample loading (*see* **Note 4**).

2. Wet the membrane by laying it on the surface of sterile distilled water and then wash briefly in 10X SSC and air-dry thoroughly.

3. Thaw the RNA samples on ice and transfer appropriate amounts (*see* **Notes 5** and **6**) to fresh tubes, add 3 vol of RNA incubation solution and mix thoroughly.

4. Heat the RNA samples at 65°C for 5 min to denature RNA secondary structure and then cool on ice.

5. Add an equal volume of ice-cold 20X SSC and mix thoroughly.

6. Dot the RNA Solution onto the membrane in 2 µL aliquots using a Gilson pipet, dry the membrane between each loading. To speed up the process it is possible to use a hairdryer to dry the membrane between loadings.

7. After loading the last sample dry the membrane and crosslink the RNA to the membrane using a UV crosslinker such as the Stratalinker (Stratagene, La Jolla, CA) (*see* **Note 7**).

8. The membrane can then be used in a hybridization experiment immediately or stored at room temperature for up to 1 mo.

4. Notes

1. To DEPC-treat solutions add 0.1% diethyl pyrocarbonate, shake to disperse and incubate at 37°C for 2 h. The solution should then be autoclaved to destroy the DEPC, which, if present, may carboxymethylate purine residues in the RNA. DEPC is highly flammable and should be handled in a fumehood, it is also suspected to be a caricinogen so should be handled with care.

2. Due to the subsequent eightfold dilution of the sample by addition of incubation solution and 10X SSC it is highly advisable to have RNA preparations at a concentration of 10 mg/mL or above. Preparations of less than 10 mg/mL can be ethanol precipitated and redissolved at higher concentrations.

3. MOPS buffer is used where formaldehyde is a component of the incubation mixture. Tris buffers are not suitable due to the reactive amine group.

4. When handling membranes, great care should be taken. Pick them up by the corners with forceps and, where possible, avoid touching the membranes even while wearing gloves.
5. The amount of RNA required for detection of expression is obviously dependent on the abundance of the transcript in the RNA population. For a cDNA clone identified by differential screening, 5 µg of total RNA per dot should be sufficient. For medium- and low-abundance transcripts it may be necessary to load more total RNA or isolate polyA$^+$ RNA for efficient detection of the transcript.
6. Where duplicate filters are being prepared, master mixes combining RNA, incubation solution and 10X SSC should be made for each RNA sample.
7. If a UV crosslinker such as the Stratalinker is not available, membranes can be crosslinked by wrapping in Saran Wrap (Dow Chemical Company, Midland, MI) and placing RNA-side-down on a UV transilluminator. However, the appropriate exposure time for crosslinking varies with the wavelength and age of the UV bulbs in the transilluminator. To establish the optimum exposure time a number of duplicate filters should be prepared and exposed for different lengths of time (20 s to 5 min). Subsequently filters should be hybridized, washed and developed together, the filter giving the strongest signal indicates the optimum exposure time for crosslinking.

References

1. Kafatos, F. C., Jones, C. W., and Efstratiadis, A. (1979) Determination of nucleic acid sequence homologies and relative concentrations by a dot hybridization procedure. *Nucleic Acid Res.* **7**, 1541–1552.

15

Nonradioactive Northern Blotting

Rainer Löw

1. Introduction

Recently, nonradioactive techniques for detection of DNA and RNA have been developed *(1,2)*. The most commonly used labels are "digoxigenin," "fluorescein," and "biotin" which are linked through a spacer to a nucleotide (in most cases UTP or dUTP) and are incorporated into specific gene-probes by various methods (e.g., random-priming, nick-translation, or PCR). The detection is based on the specific interaction of the labels with appropriate proteins, i.e., specific antidigoxigenin- and antifluorescein-antibodies or (strept)avidin, conjugated to alkaline phosphatase or peroxidase. The development of chemiluminescent substrates (e.g., CSPD, CPD, CPD-Star from Tropix) has improved the sensitivity of the detection to a range that was not achieved with colorimetric detection via chromogenic substrates (e.g., 5-chloro-4-bromo-indolylphosphate) for alkaline phosphatase.

With these tools it is now possible to detect even rare messages in total RNA samples *(3)*: 10 attomoles of a target can be detected with only one Biotin incorporated into a probe (e.g., 5'- or 3'-labeled oligonucleotide). After development of improved blotting and detection protocols the most serious problem is the quantification of transcripts on Northern blots. A linear range of signal intensity is mainly dependent on two factors: the size of the probe (and the number of incorporated labels) and the time of illuminating X-ray films. For exact quantification, the number of incorporated labels has to be calculated by comparison of the labeled probe to a dilution series of a known standard, for example 5'-labeled oligonucleotide. Subsequently, the exact amount of target in a total RNA sample can be calculated by dot- or slot-blot analysis *(4; see* Chapter 14).

From: *Methods in Molecular Biology, Vol. 86: RNA Isolation and Characterization Protocols*
Edited by: R. Rapley and D. L. Manning © Humana Press Inc., Totowa, NJ

In this chapter, we describe only one of several nonradioactive methods for Northern blot analysis using biotinylated probes. At the moment this seems to be the most sensitive method for detection of specific messages in total RNA samples. The following steps are described in detail:

1. preparation of highly labeled biotinylated-probes generated by PCR;
2. sample preparation for gel electrophoresis;
3. blotting of the electrophoretically separated RNA;
4. hybridization; and
5. detection of specifically bound probes. As starting material isolated from total RNA (for protocols; *see* **refs.** *5–7*) is necessary, it should be dissolved in formamide instead of water *(8)*.

2. Materials

All solutions and reagents should be of molecular biology-grade. If not stated otherwise, the use of distilled water throughout the blotting and detection procedures is sufficient, however, DEPC-treated distilled water is required for gel preparation. All containers and tools (e.g., forceps) should be cleaned with 1% SDS and 0.2M NaOH followed by intense watering. The wearing of gloves is recommended throughout the procedure to avoid contamination with RNases from skin.

2.1. Biotin-Labeling of DNA Probes by PCR

1. *Taq*-DNA polymerase: (5000 U/mL, Amersham International, Amersham, UK).
2. 10X *Taq*-DNA polymerase reaction buffer: 100 mM Tris-HCl, pH 9.0 (25°C), 50 mM KCl, 1.5 mM MgCl$_2$, 0.1% Triton X-100 (v/v), 0.2 mg/mL BSA.
3. 10X Biotin-16-dUTP (B-dUTP) labeling mix:

    1 mM dATP 10 µL
    1 mM dCTP 10 µL
    1 mM dGTP 10 µL
    1 mM dTTP 5 µL
    1 mM B-dUTP 5 µL

    The above labeling mix (250 µM each dNTP) with a 1:1 ratio of biotin-16-dUTP:dTTP may have to be changed when *Taq*-DNA polymerases from other suppliers are used (*see* **Note 1**). Biotin-16-dUTP can be purchased from Boehringer Mannheim (FRG). The amount of labeling mix is sufficient for eight 50 µL labeling reactions and yields approx 2 µg labeled probe (depending on probe size). The labeling mix should be stored in aliquots at –20°C to avoid multiple freeze thaw cycles (stable for 1 yr). Labeled probes can be stored at 4°C for up to 6 mo.
4. Template and oligonucleotides for the labeling reaction: Purified PCR-products or cloned (partial) cDNAs can serve as template (for standard protocols, *see* **refs.** *6, 7*). For efficient amplification between 1–1000 pg of template are sufficient. For probe amplification the specific set of primers (vector-derived or insert-specific) should be used at the highest possible stringency.

2.2. Electrophoretic Separation of Total RNA in Denaturing Agarose Gel

1. 20X MOPS buffer: $0.4M$ MOPS, $0.1M$ sodium acetate, pH 7.0, 20 mM EDTA.
2. DEPC-treated water: 0.1% diethylpyrocarbonate in double-distilled water. Stir for at least 2–3 h at room temperature (under fumehood!) and autoclave at 120°C for 45 min.
3. Formaldehyde: 37% solution ($\approx 13M$) from Sigma (St. Louis, MO). (Always handle under fumehood.)
4. Formamide: analytical grade, Sigma.
5. 10X sample buffer: 0.1% bromphenol blue, 50% glycerol; store at –20°C.
6. Loading mix (per sample): 1 μL 20X MOPS, 3.3 μL formaldehyde, 2 μL 10X sample buffer. This solution is freshly prepared (for multiple samples use mastermix).
7. Agarose: molecular biology-grade.
8. Gel chamber: Use any appropriate device. Gel chamber, gel plate, and comb are cleaned with 1% SDS, followed by $0.2M$ NaOH, and intense rinsing with distilled water.

2.3. Alkaline Blotting Procedure

1. Stock solutions:
   a. 20X SSC: $3M$ NaCl, $0.3M$ NaOAc, pH 7.0.
   b. $1M$ NaOH.
2. Transfer buffer: 5X SSC, 10 mM NaOH; pH is not adjusted and should be around 11.6. The volume needed depends on gel size and buffer reservoir. Standard gels (12 × 15 cm) can be blotted with approx 1 L of transfer buffer.
3. Neutralization buffer: 5X SSC, pH 7.0.

2.4. Additional Material

1. Sponge: soft, narrow-pore sponge, approx 2 cm thick. A thorough cleaning of the sponge before its first use is strongly recommended: Soak the sponge 2–3 times in 1% SDS followed by extended watering until SDS is removed. After transfer the sponge is well watered and dried at room temperature.
2. Whatman paper: eight pieces of Whatman No. 1 (Clifton, NJ) paper and about 2 cm paper towels cut to the size of the gel.
3. Nylon membrane: Hybond (Amersham), Tropilon (Tropix, Bedford, MA), or comparable membrane. Both neutral and positively charged membranes can be used (*see* **Note 2**).

2.5. Hybridization

1. Stock solutions:
   a. 20X SSPE: $3M$ NaCl, $0.2M$ NaH$_2$PO$_4$, 20 mM Na$_2$-EDTA, pH 7.4.
   b. $5M$ NaCl.
   c. 20% SDS (w/v).
   d. 5 μg/mL shared salmon sperm: Fragment size should be between 300–3000 bp *(6)*.
   e. Formamide: *see* **Subheading 2.2.**
   f. Polyethylene glycol 6000 (PEG, solid).

2. "Ready-to-use" solution: From the above stock solutions the prehybridization solution is prepared as follows (note that the formamide [50%] concentration may be varied according to the stringency required during hybridization): 50% formamide, 5% SDS, 6% PEG 6000 (w/v), $1 M$ NaCl, and 250 μg/mL salmon sperm. For 20 mL combine:
   a. 10 mL formamide.
   b. 5 mL SDS.
   c. 4 mL NaCl.
   d. 1 mL salmon sperm.
   e. 1.2 g PEG 6000.
   The solution is stirred under heating up to the temperature required for prehybridization. The denatured probe is added to this solution after approx 1 h of prehybridization. For denaturing, the probe is incubated in 0.5 mL of prehybrization solution at 95°C for 5 min. Probe concentration is 0.1 nM.
   Note: Salmon sperm should be denatured for 10 min at 95°C prior addition to the heated prehybridization solution.
   Note: Both prehybridization and hybridization solutions can be reused several times. The solutions are stable for 6 mo at 4°C. Before use the solutions should be denatured again at 85°C and cooled down to approx 50°C before addition to the blot.
3. Wash buffers (after hybridization):
   a. Low stringency wash: 2X SSPE, 0.5% SDS (buffer #1).
   b. High stringency wash: 0.2–0.4X SSPE, 0.5% SDS (buffer #2).

2.6. Detection

1. Stock solutions:
   a. 10X PBS: $0.4 M$ Na$_2$HPO$_4$, $0.1 M$ NaH$_2$PO$_4$, $1 M$ NaCl, pH 7.4.
   b. 20% SDS (w/v).
   c. Casein (alkaline phosphatase, RNase- and DNase-free, Tropix).
   d. Substrate and avidin (streptavidin)-alkaline phosphatase conjugate are included in the "Southern light"-kit (Tropix).
   e. X-ray film: Kodak XOMAT (Kodak, Rochester, NY), Hyperfilm (Amersham) or any other film appropriate for chemiluminescent detection.
2. Blocking buffer: 1X PBS, 0.5% SDS, 0.1–0.2% casein (w/v, alkaline phosphatase free); has to be freshly prepared. Use a microwave oven or heating block to heat the PBS/casein mixture to 70°C (500 W, 45 s) for dissolving casein while stirring for 10 min. Before use, the hot solution is adjusted to room temperature. If casein is not completely dissolved, filter the solution.
3. Conjugation buffer: 1:6000 dilution of AVIDx-alkaline phosphatase conjugate (Tropix) in blocking buffer (has to be freshly prepared).
4. Wash buffer: 1X PBS, 0.5% SDS.
5. Assay buffer: $0.1 M$ diethanolamine, pH 10.0, 1 mM MgCl$_2$. The solution is stable at 4°C for about 1 wk when 1 mM NaN$_3$ is added.
6. Substrate buffer: 1:100 dilution of CSPD (25 mM stock, Tropix) in assay buffer (has to be freshly prepared and is stable for several hours at 4°C).

3. Methods

3.1 Biotin-Labeling of DNA-Probes by PCR

When biotin-labeled probes are synthesized by PCR, the annealing temperature and extension time will depend on primer sequence(s) and product length, respectively. For illustration, a representative example is given in which an insert ligated into a standard vector (pBluescript SK+ vector; Stratagene, La Jolla, CA) is used as template for synthesis of a biotinylated probe. The vector-based primers T3 and T7 bind to the left and right from the Eco RV site (within the multiple cloning site). In principle, with this primer set probes can be amplified for any insert; however, controls should be run to exclude that a chance priming occurs within the insert of interest.

1. For a 50 µL reaction combine at room temperature:
   a. 35 µL Template (1 pg–1 ng) and water.
   b. 5 µL 10X reaction buffer.
   c. 5 µL 10X bio-dNTP-mix.
   d. 2.5 µL sense primer (10 pmol/µL).
   e. 2.5 µL antisense primer (10 pmol/µL).
2. The PCR program used to amplify probes of up to 700 bp is as follows:
   a. 94°C–5 min (1X).
   b. 94°C–30 s, 55°C–60 s, 72°C–30 s (30X).
   c. 72°C–5 min (1X).

3.2 Electrophoretic Separation of Total RNA in a Denaturing Agarose Gel

After isolation of total RNA according to standard protocols *(5–7)* the final RNA-pellet is dissolved in formamide *(8)* to a concentration of 1–2 µg/µL.

1. To 78.3 mL distilled water (DEPC-treated) add 1.5 g agarose to a 200 mL screw-capped bottle. After melting the agarose in a microwave oven, add 5 mL 20X MOPS and 16.6 mL formaldehyde. After intense mixing, cast the gel immediately (under a fumehood) and allow to solidify at room temperature.
2. Prior to loading, dilute the samples to 13.7 µL in formamide. Add 6.3 µL of loading mix to the samples and denature for 5–10 min at 65°C and cooled in ice-water for additional 5 min.
3. Electrophoretic separation at 5–7.5 V/cm gel length in 1X MOPS, 1% formaldehyde (v/v). Separation time depends on the resolution required (approx 2–3 h).

3.3. Alkaline Blotting Procedure

1. After electrophoresis soak the gel twice for 10 min in transfer buffer. The assembly for transfer is as depicted in **Fig. 1**.
2. Place the transfer buffer-soaked sponge in the middle of a clean container filled with transfer buffer. Care should be taken to remove all air bubbles from the

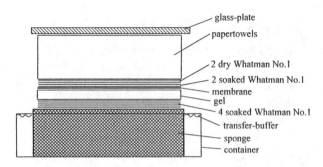

Fig. 1. Assembly for sponge-based capillary transfer of nucleic acids.

sponge surface. On top of the sponge four layers of transfer buffer-soaked Whatman No. 1 blotting paper are placed, followed by the gel (upside down) and the membrane. On top of the membrane place two soaked and two dry Whatman No. 1 blotting papers, and approx 2 cm of paper towels.

3. The transfer "sandwich" is fixed by a weight (e.g., glassplate) placed on top. *Note:* To avoid excessive compression of the gel matrix, the weight should not exceed 2–3 g/cm^2 for gels of 1–2%. Furthermore, transfer time is very critical and should not exceed 10 min/mm of gel thickness (*see* **Note 4**).

4. After transfer is completed rinse the membrane for 5 min in neutralization buffer and air-dry on Whatman paper for 2 min before it is UV crosslinked. Before starting the prehybridization, the blot has to be dried completely for at least 2 h at room temperature (*see* **Note 5**).

3.4. Hybridization

Prehybridization and hybridization are performed in a commercial hybridization oven or a waterbath at the appropriate temperature. Hybridization temperature is dependent on salt concentration, formamide concentration, GC-content, and degree of homology between probe and target RNA. Usually temperature is in the range from 37–42°C. Empirical formulas are available to calculate the stringency for specific hybrid formation (*see* **Note 6**).

1. Heat the hybridization solution to the required temperature.
2. Place the blot into the container or tube and pour the prewarmed hybridization solution over the blot.
3. After 1 h, add the denatured probe directly into the prehybridization solution. Hybridization is usually carried out overnight at the same temperature as the prehybridization.
4. After overnight hybridization, wash the blot twice at low stringency in buffer #1 for 10 min at room temperature. The final stringency is determined via the high stringency wash in buffer #2 at 55–65°C for 30–60 min. Carry out the washes under gentle shaking (80–120 rpm). After the high stringency wash, the blot is ready for the detection procedure.

3.5. Detection

We present an optimized protocol for the detection of low- to medium-abundant mRNAs in samples of total RNA. For the detection of specific transcripts in poly(A^+) enriched RNA preparations or abundant transcripts in total RNA modifications of the protocol are required (*see* **Note 7**). All incubations should be carried out on a rotary shaker at low rotation (30–80 rpm/min) at room temperature.

1. Incubate the blot in blocking-buffer twice for 5 min (80 rpm) and once for 30 min (30 rpm). Note that excessive blocking reduces the final signal strength but may improve the signal-to-noise ratio.
2. Drain the blocking buffer and add the conjugation buffer to the blot. Conjugation is performed for 1 h (30 rpm). Note that addition of conjugation buffer should be done at one corner of the container and not over the blot surface.
3. After conjugation transfer the blot to a new container. Allow excess conjugation buffer to drain from the blot surface. Following two washes with blocking buffer (5 min, 80 rpm) the membrane is washed four times with wash buffer (5 min, 80 rpm).
4. Equilibration to appropriate reaction conditions for alkaline phosphatase is achieved through two washes in assay buffer for 5 min (80 rpm). Note that it is essential to remove excess assay buffer before starting the incubation with substrate buffer to avoid dilution of substrate. This is best done by placing the blot upside-up for 2 min on two layers of assay buffer-wetted Whatman No. 1 paper.
5. While the blot is laying on moist Whatman paper, add substrate buffer to the empty container. Subsequently, place the blot upside-down on the liquid film of substrate buffer and incubate for 10 min (30 rpm). Finally, remove excess substrate buffer from the membrane as described in **step 4**.
6. For film exposure, place the moist membrane between two plastic covers (e.g., Cooking-foil, Melitta, *see* **Note 8**). Transfer the covered membrane to a film exposure cassette and expose to the X-ray film for 1–2 h.

3.5.1. Flow-Sheet for the Complete Method

Time (h)	Procedure
3	Electrophoretic separation (include gel casting and running time)
3.5	Alkaline blotting (includes incubation in transfer buffer, transfer and drying after transfer)
1.5	Prehybridization (includes preparation of a fresh solution)
12–18	Hybridization
4–4.5	Detection

3.5.2. Detection Procedure

Time (min)	Procedure
20	Low stringency wash (2 × 10 min)
30–60	High stringency wash
40	Blocking (includes 2 × 5 min and 1 × 30 min incubations)
60	Conjugation
30	Washes (includes 2 × 5 min in blocking buffer and 4 times in wash buffer)
10	Equilibration (2 × 5 min in assay buffer)
15	Substrate incubation (includes removing of excess buffer before and after incubation)
60–120	Film-exposure

4. Notes

1. Using a different *Taq*-DNA-polymerase may require an optimization of label incorporation because of enzyme-specific discrimination between dTTP and bio-dUTP. This is done by changing the dTTP:bio-dUTP ratio while keeping the remaining dNTP mixture constant. Mostly, a dTTP:bio-dUTP of 1:1 ratio will be sufficient, and this holds also for labels other than biotin. The sensitivity of the probe has to be determined by slot-blot before using it for northern blots. For successful northern blot detection of low-abundant mRNAs 1 pg (or below) of target should be detected.

2. The type of membrane used for transfer does not seem to be critical, however, from our experience the use of neutral nylon membranes yields a better signal-to-noise ratio, especially after longer film exposures (e.g., above 2 h).

3. Probe concentration is not critical within the range of 0.5–2.0 nM. It should be noted that higher probe concentrations allow one to shorten the hybridization time. However, this will also lead to higher background. Probe concentrations lower than indicated will need longer hybridization times to achieve equal signal strength.

4. For yet unknown reason(s) transfer time is very critical. For example, after three hours of blotting, signals, which were easily detectable after one hour of blotting, become very weak or not detectable. This is not due to nonspecific RNA hydrolysis (resulting from alkaline conditions during transfer), because the abundant rRNAs are detectable with methylene blue staining on the membrane.

5. Proper blot drying after UV crosslinking increases signal strength. Furthermore, the background is significantly reduced.

6. Calculation of stringency follows empirical formulas taking into account the different hybrid stabilities of DNA–DNA and DNA–RNA hybrid molecules:

DNA–DNA Hybrid *(6)*

$$T_m = 81.5 + 16.6*\log[Na^+] + 0.4*(\%GC) - 0.6*(\%formamide) - (600/N)$$

DNA–RNA Hybrid *(9)*

$$T_m = 79.8 + 18.5 \log [Na^+] + 0.58 (\%GC) + 11.8$$
$$(\%GC)^2 - 0.35 (\%formamide) - (820/L)$$

where T_m = melting point of hybrid; %GC= percent of GC in DNA molecule; N/L = length of the hybrid molecule in bp; $[Na^+]$ = molar concentration of Na^+; %formamide = percent of formamide in the hybridization solution.

7. Target abundance determines the sensitivity required for detection. For high-abundant mRNAs hybridization time may be reduced. The detection procedure can also be shortened: Conjugation could be done in 20–30 min, and a substrate incubation of 5 min is sufficient. For 18S rRNA film exposure requires less than 10 min!

8. The plastic foil for film exposure has to meet two requirements: First, it should not absorb at the emission wavelength of the chemiluminescent substrate, and, second, it should not be affected by the substrate buffer. Both points should be checked before use.

9. Strong background caused by nonspecific binding of probe can be reduced via RNase A-treatment after hybridization. Note that the blot is no longer useful for further hybridizations. Alternatively, hybridization and washing stringencies may be modified.

10. Reasons for a spotty background:
    a. the AP-conjugate was not centrifuged before dilution into blocking buffer.
    b. the AP-conjugate is too old.
    c. the temperature during conjugation was too high.

11. Reasons for a cloudy background:
    a. the casein may be too old.
    b. the blocking buffer was too hot during casein dissolution.
    c. temperature during film exposure was above 30°C.

12. The same blotting and detection procedures are routinely used in our lab for Southern blot analysis *(4)*. Stringency conditions required during hybridization are calculated according to the formula given in **Note 6.**

References

1. Lanzillo, J. J. (1991) Chemiluminescent nucleic acid detection with digoxigenin-labeled probes: a model system with probes for angiotensin converting enzyme which detect less than one attomole of target DNA. *Anal. Biochem.* **194**, 45–53.
2. Klevan, L. and Gebeyehu, G. (1990) Biotinylated nucleotides for labelling and detection of DNA. *Methods Enzymol.* **184**, 561–577.
3. Löw, R. and Rausch, T. (1994) Sensitive non-radioactive northern blots using alkaline transfer of total RNA and PCR amplified biotinylated probes. *BioTechniques* **17**, 1026–1030.
4. Löw, R. and Rausch, T. (1996) Non-radioactive detection of nucleic acids with biotinylated probes, in: *A Laboratory Guide to Biotin-Labeling in Protein and Nucleic Acid Analysis.* (T. Meier and F. Fahrenholz, eds.), Birkhäuser Verlag, Basel, pp. 201–213.

5. Logemann, J., Schell, J., and Willmitzer, L. (1987) Improved method for the isolation of RNA from plant tissues. *Anal. Biochem.* **163,** 16–20.
6. Maniatis, T., Fritsch, E. F., and Sambrook, J. (1989) *Molecular Cloning: A Laboratory Manual,* Cold Spring Harbor Laboratory, Cold Spring Harbor, NY.
7. Ausubel, F., Brent, R., Kingston, R. E., Moore, D. D., Seidman, J. G., Smith, J. A., and Struhl, K. (1995) *Current Protocols in Molecular Biology* (vol. 1, Chapter 4). Wiley, Interscience, NY.
8. Chomczynski, P. (1992) Solubilization in formamide protects RNA from degradation. *Nucleic Acids Res.* **28,** 3791.
9. Casey, J. and Davidson, N. (1977) Rates of formation and thermal stability of RNA:RNA and DNA:DNA duplexes at high concentrations of formamide. *Nucleic Acids Res.* **4,** 1539–1552.

16

The Use of RNA Probes for the Analysis of Gene Expression

Northern Blot Hybridization and Ribonuclease Protection Assay

Dominique Belin

1. Introduction

The isolation and characterization of RNA polymerases from the *Salmonella* phage SP6 and the *E. coli* phages T7 and T3 has revolutionized all aspects of the study of RNA metabolism *(1–6)*. Indeed, it is now possible to generate unlimited quantities of virtually any RNA molecule in a chemically pure form. This technology is based on a number of properties of the viral transcription units. First, and in contrast to their cellular counterparts, the enzymes are single-chain proteins which were easily purified from phage-infected cells and are now produced by recombinant DNA technology. Second, they very specifically recognize their own promoters (*7* and references therein), which are contiguous 17–20 bp long sequences rarely encountered in bacterial, plasmid or eukaryotic sequences. Third, the enzymes are highly processive, allowing the efficient synthesis of very long transcripts from DNA templates. In this chapter, the author will discuss the preparation of the DNA templates, the transcription from the templates of labeled synthetic RNA molecules, commonly called riboprobes, and their use in Northern and RNase protection assays.

2. Materials

These protocols require the use of standard molecular biology materials and methods for carrying out subcloning, PCR and gel electrophoresis, in addition to those listed below. More rigorous precautions are required for working with RNA are than commonly used for most DNA studies, because it is important to avoid RNase contamination. The two major sources of unwanted RNases are

From: *Methods in Molecular Biology, Vol. 86: RNA Isolation and Characterization Protocols*
Edited by: R. Rapley and D. L. Manning © Humana Press Inc., Totowa, NJ

the skin of investigators and microbial contamination of solutions. Most RNases do not require divalent cations and are not irreversibly denatured by autoclaving. Gloves should be worn and frequently changed. Sterile plastic ware should be used, although some mechanically manufactured tubes and pipet tips have been used successfully without sterilization. Glassware should be incubated at 180°C in a dried baking oven for several hours.

In addition, divalent cations (such as Mg^{2+} or Ca^{2+}) accelerate base-mediated RNA hydrolysis. RNA can be stored in sterile water but is most stable when stored in 1 mM KOAc pH 5.0, 0.1 mM EDTA.

2.1. Preparation of Riboprobes

1. Water: while a number of protocols recommend treatment of the water used for all solutions with diethylpyrocarbonate, I find this to be unnecessary. Double-distilled water is used for the preparation of stock solutions that can be autoclaved (121°C, 15–30 min), and sterilized water is used otherwise.
2. TE: 10 mM Tris-HCl, pH 8.1, 1 mM EDTA.
3. 10X TB: 0.4M Tris-HCl, pH 7.4, 0.2M NaCl, 60 mM MgCl$_2$, 20 mM spermidine. If ribonucleotides are used at concentrations > 0.5 mM each, MgCl$_2$ concentration should be increased to provide a free magnesium concentration of 4 mM.
4. 0.2M DTT: The solution is stored in small aliquots at –20°C. Aliquots are used only once. EDTA can be included at 0.5 mM to stabilize DTT solutions.
5. Ribonucleotides: Neutralized solutions of ribonucleotides are commercially available, or can be made up from dry powder (*see* **Note 1**). Ribonucleotide solutions can be stored at –20°C for a few months.
6. RNA polymerase stocks: The three RNA polymerases, SP6, T3, and T7, are available commercially; store at –20°C.
7. RNA polymerase dilution buffer: 50 mM Tris-HCl, pH 8.1, 1 mM DTT, 0.1 mM EDTA, 500 µg/mL BSA, 5% glycerol. Diluted enzyme is unstable and should be stored on ice for no more than a few hours.
8. Stop-mix: 1% SDS, 10 mM EDTA, and 1 mg/mL tRNA. The tRNA may be omitted.
9. TEN buffer: 10 mM Tris-HCl, pH 8.1, 1 mM EDTA, 100 mM NaCl.
10. Sample buffer (polyacrylamide/urea gels): 80% deionized formamide (*see* **Note 2**), 2M urea, 0.1X TBE (8.9 mM Tris, 8.9 mM boric acid, 0.2 mM EDTA), and 0.01% each of xylene cyanol and bromophenol blue. Use 1–2 µL per µL of RNA in aqueous solution.

2.2. Northern Blot Hybridization

1. Formamide: Pure formamide is slowly hydrolyzed by water vapor to ammonium formate and therefore must be deionized (*see* **Note 2**). Store at –20°C.
2. Glyoxal: a 30% gyloxal solution (6M) is deionized by several incubations at room temperature with a mixed bed resin (AG501-X8, BioRad), until the pH is 6–7 and the conductivity of 3% glyoxal in water is below 30 µSiemens. Store at –20°C in small aliquots.

3. Denaturation Buffer: 1M glyoxal in 50% DMSO and 10 mM Na$_2$HPO$_4$, pH 6.8. Ethidium bromide can be included at a concentration of 50 μg/mL to visualize the rRNAs during electrophoresis.

4. 5X RSB: 50% glycerol, 10 mM Na$_2$HPO$_4$, pH 6.8, 0.4% bromophenol blue. Autoclave and store at –20°C in small aliquots.

5. Transfer membrane: Nitrocellulose (Schleicher and Schuell, Keene, NH) and nylon membranes (Hybond N, Amersham, Arlington Heights, IL or Biodyne A, Pall, Glen Cove, NY) have been used successfully.

6. 50X Denhardt's solution: Dissolve 2 g of bovine serum albumin (fraction V) in 80 mL of sterile water. Bring the pH to 3.0 with 2N HCl, boil for 15 min and cool on ice for 10 min. Bring the pH to 7.0 with 2N NaOH and add sterile water to 100 mL. Autoclave a 100 mL solution of 2% polyvinylpyrrolidone (K90, Fluka, Buchs, Switzerland), and 2% ficoll 400 (Pharmacia, Piscataway, NJ), and add to the 2% albumin solution.

7. Hybridization solution: 50% deionized formamide, 0.8M NaCl, 50 mM Na-PIPES, pH 6.8, 2 mM EDTA, 0.1% SDS, 2.5X Denhardt's solution, and 0.1 mg/mL sonicated and heat-denatured salmon sperm DNA. The hybridization solution is heated, filtered over 0.45 μm Nalgene sterilization units or Millipore nitrocellulose filters, and kept at the hybridization temperature during prehybridization. The filtration step decreases non-specific attachment of the probe to the membrane and degasses the solution.

8. 20X SSC : 3M NaCl, 0.3M Na/citrate, pH 7.0.

2.3. RNase Protection

1. Hybridization mixture for RNase protection: 80% deionized formamide (*see* **Note 2**), 0.4M NaCl, 40 mM Na-PIPES, pH 6.8, and 1 mM EDTA.

2. RNase digestion buffer: 300 mM NaCl, 10 mM Tris-HCl, pH 7.4, 4 mM EDTA.

3. Pancreatic RNase: Make a 10 mg/mL solution in TE containing 10 mM NaCl. Vials containing lyophilized RNases should be carefully open in a ventilated hood to avoid contamination. Boil for 15 min and slowly cool to room temperature. Store in aliquots at –20°C.

4. T1 RNase: make a 1 mg/mL solution in TE. Adjust the pH to 7.0. Store in aliquots at –20°C.

5. Proteinase K: dissolve the enzyme at 20 mg/mL in water and store in aliquots at –20°C.

3. Methods

3.1. Preparation of Riboprobes

3.1.1. Linearized Plasmid Templates for Runoff Transcription

1. Subclone the desired gene fragment in a transcription vector (*see* **Note 3**).

2. Isolate plasmid DNA by alkaline lysis from a 30–100 mL saturated culture. Plasmid DNAs are purified by CsCl/ethidium bromide centrifugation or by precipitation with polyethylene glycol (*see* **Note 4**).

3. Linearize 2–20 μg of plasmid DNA with an appropriate restriction enzyme (*see* **Note 5**). Verify the extent of digestion by electrophoresis of an aliquot (0.2–0.5 μg

of DNA) on an agarose minigel in the presence of ethidium bromide (0.5 µg/mL) (*see* **Note 6**).

4. Purify the restricted DNA by two extractions with phenol:chloroform:isoamyl alcohol (25:24:1 by vol), and remove residual phenol by one extraction with chloroform (24/1 by vol). Precipitate the DNA with ethanol. If little DNA is present (below 5 µg), add 10 µg of glycogen (Boehringer, Mannheim, Germany) as a carrier; this carrier has no adverse effect in the transcription reactions. After washing the ethanol pellet, the DNA is dried in air and resuspended in TE at 1 µg/ mL.

3.1.2. Synthetic and PCR-Derived Templates

The major limitation in using restriction enzymes to clone inserts and to linearize plasmid templates is that appropriate sites are not always available. Furthermore, the transcripts will almost always contain 5' and 3' portions that differ from those of endogenous RNA. One possibility to circumvent these difficulties is based on the transcription of small DNA fragments obtained by annealing of synthetic oligodeoxynucleotides *(6,8)*. An alternative and more general approach generates the transcription templates via PCR amplification of plasmid DNA. In theory, such templates could direct the synthesis of virtually any RNA sequence.

1. Design the 5' primer, which has a composite sequence: its 5' portion is constituted by a minimal T7 promoter, and its 3' portion corresponds to the beginning of the transcript (*see* **Note 7**). Six to ten nucleotides are usually sufficient to prime DNA synthesis on the plasmid template.
2. Design the 3' primer, which is usually 17–20 nt long and defines the 3' end of the transcript (*see* **Note 8**).
3. PCR amplify 2–50 ng of plasmid DNA in a total volume of 100 µL. Verify that a DNA fragment of the expected size has been amplified, and estimate the amount of DNA by comparison with known standards.
4. Purify the DNA as described above and resuspend in TE at the appropriate concentration. The PCR-derived templates are transcribed at lower DNA concentrations than plasmids to maintain the molar ratio of enzyme to promoter; I use 3 µg/mL for a 100 bp fragment.

3.1.3. Basic Transcription Protocol for Radioactive Probes

1. Assemble the transcription mixture to a total volume of 10 µL by adding in the following order: water (as required), 1 µL of 10X TB, 0.5 µL of bovine serum albumin (2 mg/mL), 0.5 µL of 0.2M DTT, 0.25 µL of placental RNase inhibitor (40 U/µL), 1 µL of a 5 mM solution of each ribonucleotide (i.e., ATP, GTP, CTP), 5 µL of α-[^{32}P]-UTP (400 Ci/mmole, 10mCi/mL), 1 µL of restricted plasmid DNA template and 0.25–0.75 µL of RNA polymerase (*see* **Notes 9** and **10**).
2. Incubate 40 min at 37°C for T7 and T3 polymerases or 40°C for SP6 polymerase. After addition of the same number of units of enzyme (*see* **Note 10**), incubate for a further 40 min.

3. Degrade the template with RNase-free DNase (1 U/μg of DNA) for 20 min at 37°C.
4. Stop the reaction by adding 40 μL of stop-mix.
5. After two extractions with phenol/chloroform, in which the organic phases are back-extracted with 50 μL of TEN, the combined aqueous phases are purified from unincorporated nucleotides by spun-column centrifugation. The spun-column is prepared by filling disposable columns (QS-Q, Isolab, Hörth, Germany) with a sterile 50% slurry of G-50 Sephadex (Pharmacia, Piscataway, NJ) in TEN followed by centrifugation for 5 min at 200g. The samples are carefully deposited on top of the dried resin, and the column is placed in a sterile conical centrifuge tube. After one centrifugation for 5 min at 200g, 200 μL of TEN is deposited on top of the resin and the column recentrifuged. The eluted RNA (200–400 μL) is ethanol precipitated and resuspended in water (*see* **Note 11**).
6. Measure the incorporation efficiency by counting an aliquot (*see* **Note 12** for calculations and for modifications of the basic protocol).
7. The size of the transcript may be verified by electrophoresis in polyacrylamide/urea gels (*see* **Note 13**).

3.2. Northern Blot Hybridization

The use of riboprobes in RNA blot hybridizations follows the same general principles as that for DNA probes. The major disadvantage in using double-stranded DNA probes results from self-annealing, which decreases the availability of DNA probes to bind to the immobilized target; this is particularly critical with heterologous probes, where the reannealed probe may displace incompletely matched hybrids. Self-annealing of course does not occur with single stranded RNA probes. Most difficulties encountered with riboprobes stem from the increased thermal stability of RNA:RNA hybrids. Thus, cross hybridization of GC-rich probes to rRNAs can generate unacceptable backgrounds. This problem is often solved by increasing the stringency of hybridization, as illustrated in **Fig. 1B**. Alternatively, the template may have to be shortened to remove GC-rich regions from the probe.

1. Denature the sample in 8 μL of denaturation buffer for 15–30 min at 50°C. Add 2 μL of 5X RSB and electrophorese in 0.7–2% agarose gels in 10 mM Na$_2$HPO$_4$, pH 6.8. The buffer should be circularized with a peristaltic pump, so that the pH near the electrodes remains neutral.
2. Transfer the RNAs by capillary transfer onto a membrane (*see* **Note 14**).
3. Fix the RNA by incubating the blots at 80°C under vacuum. This step is essential to remove glyoxal covalently fixed to guanine residues in the RNA.
4. UV-crosslinking is often used to improve RNA retention, although irradiation may increase background hybridization (*see* **Fig. 1A** and **Note 15**).
5. Prehybridize the blots for 4–12 h in hybridization solution. Use 200 μL/cm^2 of membrane.

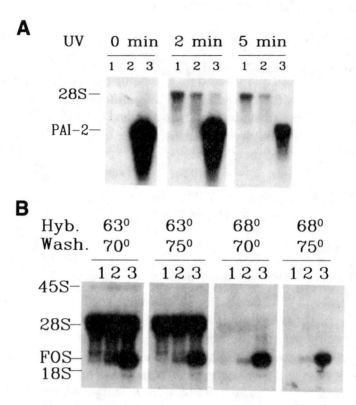

Fig. 1. Northern blot hybridization with riboprobes. **(A)** Effect of UV cross-linking. Northern blot hybridization of PAI-2 mRNA in murine total cellular RNA with an homologous cRNA probe. Lanes 1 and 2: 5 µg of placental RNA (15.5 and 18.5 d of gestation), that do not contain detectable levels of PAI-2 mRNA. Lane 3: 1 µg of LPS-induced macrophage RNA, an abundant source of PAI-2 mRNA *(29)*. All samples were electrophoresed and transferred together. After cutting the membrane, each filter was UV-treated as described. The filters were hybridized at 58°C, washed at 70°C, and exposed together. Cross-hybridization of the probe to 28S rRNA is more pronounced after UV irradiation, and specific hybridization is decreased after 5 min of UV exposure. **(B)** Effect of hybridization temperature. Northern blot hybridization of c-*fos* mRNA in rat total cellular RNA with a murine v-*fos* cRNA probe. Lanes 1: uninduced cells. Lanes 2: partially induced cells. Lanes 3: fully induced cells (M. Prentki and D. Belin, unpublished). All samples were electrophoresed and transferred together. The filters, which were not UV cross-linked, were hybridized and washed in parallel at the indicated temperatures . The four filters were exposed together. Cross-hybridization of the probe to 28S rRNA was essentially abolished by hybridizing at 68°C. Some specific signal was lost with the 75°C stringency wash.

6. Dilute the probe in hybridization solution (25–50 µL/cm² of membrane), then add to the membrane. I frequently hybridize 2–3 filters per bag. Hybridize at the

appropriate temperature (58–68°C, see **Fig. 1B** and **Note 16**), usually for 12–18 h (*see* **Note 17**).

7. Wash the membranes twice for 10–20 min at the hybridization temperature with 100 μL/cm² of 3X SSC and 2X Denhardt's solution, and then three times with 0.2X SSC, 0.1% SDS, and 0.1% Na pyrophosphate at the appropriate temperature (*see* **Fig. 1B** and **Note 17**).

8. Expose to autoradiographic film. As long as the membranes are not allowed to dry, they can be further washed at increased stringencies to reduce background.

3.3. RNase Protection

This assay is based on solution hybridization and on the resistance of RNA:RNA hybrids against single strand specific RNases. A ^{32}P-labeled probe is synthesized that is partially complementary to a portion of the target RNA. It is hybridized in excess to the target so that all complementary sequences are driven into the labeled RNA:RNA hybrid. Unhybridized probe and any single-stranded region of the hybridized probe are then removed by RNase digestion. The "protected" probe is then detected and quantitated on a denaturing polyacrylamide gel. It can be used to map the ends of RNA molecules or exon-intron boundaries. It also provides an attractive and highly sensitive alternative to Northern blot hybridization for the quantitative determination of mRNA abundance.

RNase protection has a number of advantages. First, solution hybridization tolerates high RNA input (up to 60 μg of total RNA), and is not affected by the efficiency of transfer on membranes or by the availability of membrane-bound RNAs. Second, the signal to noise ratio is much more favorable, since cross-hybridizing RNAs yield only short protected fragments. Third, a significant fraction of mRNAs is often partially degraded during RNA isolation; in Northern blots, this generates a trail of shorter hybridizing species, which reduces the sensitivity of detection. Finally, the detection of hybridized probes on sequencing gels is much more sensitive because the width of the bands is less than a tenth of those of intact RNAs in agarose gels. Only two features of Northern blots are lost in RNase protection assays: complete size determination of target RNAs and multiple use of each sample.

1. Linearize the plasmid DNA template as described in **Subheading 3.1.1.** (*see* **Note 18**).

2. Transcribe the template as described in **Subheading 3.1.3.** The amount of labeled ribonucleotide may be varied (*see* **Note 19**).

3. An optional step is to purify the full length transcripts by electrophoresis (*see* **Note 20**). Separate the transcript on a preparative 5–6% polyacrylamide/urea gel (gel thickness: 0.4–1.5 mm). Cover the wet gel within Saran Wrap™ and expose for 30 s to 5 min at room temperature to localize the full length transcript. Cut the exposed band on the film with a razor blade. After aligning the cut film on the gel, excise the gel band with a sterile blade. The cut gel should be re-exposed to verify that the correct band has been excised.

4. Elute the RNA from the gel either by diffusion or electroelution.

   a. Incubate the gel fragment in an Eppendorf tube in 500 μL of 0.5*M* ammonium acetate, 1% SDS and 20 μg/mL tRNA for 1–3 h at 37°C, or overnight at 4°C. The eluate and residual gel can be counted to ensure that more than 60% of the RNA is eluted. After two extractions with phenol/ chloroform, recover the eluted RNA by ethanol precipitation.

   b. Electroelute for 1–2 h at 30 V/cm in 0.1X TBE in a sterile dialysis bag, after which invert the polarity for 30 s, to detach the eluted RNA from the membrane. Purify the eluate by two phenol/chloroform extractions and ethanol precipitation with an known amount of tRNA carrier. This procedure is very sensitive to RNase degradation.

5. Resuspend the probe in water at 1–2 ng/ μL. Add 1 μL of probe to 29 μL of hybridization mixture for each assay. The exact amount of probe is not critical, since it is in excess over its specific target (*see* **Notes 19** and **21**).

6. Lyophilize or ethanol precipitate the sample RNAs (*see* **Note 22**). Resuspend in 30 μL of complete hybridization mixture including probe, heat for 2 min at 90°C, and incubate overnight, usually at 45°C (*see* **Notes 23** and **24**).

7. Cool the samples on ice and add 300 μL of RNase digestion buffer. Digest for 1 h at 25°C with pancreatic RNase, which cleaves after uracil and cytosine residues, with T1 RNase, which cleaves after guanine residues, or with both RNases (*see* **Notes 25** and **26**).

8. Add 20 μL of 10% SDS, and degrade the enzyme(s) with 0.5 μL (10 μg) of proteinase K for 10–20 min at 37°C. Extract twice with phenol/chloroform, and precipitate the RNAs with ethanol with 10 μg of carrier tRNA (*see* **Note 27**).

9. Resuspend the RNAs in sample buffer, denature the hybrids for 2 min at 90°C, and electrophorese in polyacrylamide/urea sequencing gels. Alternatively, the hybrids may be analyzed on non-denaturing polyacrylamide gels. Fix the gels with 20% ethanol and 10% acetic acid to remove the urea, dry, and autoradiograph (*see* **Note 28**).

4. Notes

1. Powdered ribonucleotides should be resuspended in water, neutralized to pH 7.0 with 1*M* NaOH or HCl, and adjusted to the desired concentration by measuring the UV absorbance of appropriate dilutions :

   100 m*M* ATP: 1540 absorbance units at 259 nm
   100 m*M* GTP: 1370 absorbance units at 253 nm
   100 m*M* CTP: 910 absorbance units at 271 nm
   100 mM UTP: 1000 absorbance units at 262 nm

   The integrity of ribonucleotide triphosphate solutions can be verified by thin layer chromatography on PEI-cellulose (PEI-CEL300). The resin is first washed with water by ascending chromatography to remove residual UV-absorbing material and dried. 10–30 nmoles of ribonucleotides are deposited on the resin, which is then resolved by ascending chromatography with 0.5*M* KH_2PO_4, adjusted to pH 3.5 with H_3PO_4. After drying, the ribonucleotides are detected by UV-shadowing at 254 nm.

2. To deionize formamide, incubate at –80°C until 75–90% of the solution has crystallized. Discard the liquid phase, thaw, and incubate at 4°C for several hours with a mixed bed resin (AG501-X8, BioRad). Use a Teflon™-covered magnet which has been freed of RNase by treatment with 0.1M NaOH for 10 min, and rinsed with water and crude formamide. Check the conductivity, which should be below 20 μSiemens. Filter the solution onto sterile paper (Whatman, LS-14) over a sterile funnel. The absorbance at 270 nm should be below 0.2.

3. All the vectors that are commercially available consist of high copy number *E.coli* plasmids derived from ColE1. The original plasmids (pSP64 and pSP65, Promega, Madison WI) contained only one SP6 promoter located upstream of multiple cloning sites (MCS) *(2)*. In the second generation of plasmids, two promoters in opposite orientation flank the MCS to allow transcription of both strands of inserted DNA fragments. The pGEM™ series (Promega) contain SP6 and T7 promoters *(5)*, while the pBluescript™ series (Stratagene, La Jolla, CA) contain T7 and T3 promoters.

   The choice of plasmid is mostly a matter of personal preference, although it can be influenced by the properties of individual sequences. For instance, we have frequently observed premature termination with SP6 transcripts. The nature of the termination signals is not completely understood *(4,9)*, and their efficiency can be more pronounced when one ribonucleotide is present at suboptimal concentration (*see* **Note 13**). The problem has been sometimes solved by recloning the inserts in front of a T7 or T3 promoter. The partial recognition of T7 (T3) promoters by T3 (T7) polymerases may result in the transcription of both strands when the ratio of enzyme to promoter is not carefully controlled. This can be a source of artifacts, particularly when the templates are linearized inside the cloned inserts.

4. It is possible to use plasmid DNA from "minipreps," although transcription efficiency can be reduced, particularly with SP6 polymerase. The RNA present in the "minipreps" is digested with pancreatic RNase (20 μg/mL), which is removed during purification of the linearized templates. Spun-column centrifugation of the digested "minipreps" can improve transcription efficiency.

5. Since RNA polymerases can initiate transcription unspecifically from 3' protruding ends *(4,10)*, restriction enzymes that generate 5' protruding or blunt ends are usually preferred. If the only available site generates 3' protruding ends, the DNA can be blunt-ended by exonucleolytic digestion with T4 DNA polymerase or with the Klenow fragment of DNA polymerase I. Restriction with enzymes which cut the plasmids more than once may also be used, provided that the promoter is not separated from the insert.

6. The plasmids must be linearized as extensively as possible. Since circular plasmids are efficient templates, their transcription by the highly processive enzymes may yield RNA molecules that can be up to 20 kb long and thus incorporate a significant portion of the limiting ribonucleotide.

7. There are constraints on the 5' sequence of transcripts since the sequence immediately downstream of the start site is necessary for the transition from an

abortive initiating cycling mode to the elongation mode. The first 6 nt have a strong influence on promoter efficiency; in particular, the presence of uracil residues are usually detrimental *(6,8)*. It may be necessary, therefore, to include in the 5' end of the transcripts 5–6 bases that differ from those present in natural RNAs. My colleagues and I have used a number of composite T7 promoters, whose efficiency is summarized below. In addition to the promoter sequence (–17 to –1:5'-TAATACGACTCACTATA) at the 5' end, the first 6 nt of the templates are:

Efficient promoters	Inefficient promoters
GGGAGA (T7 consensus)	GTTGGG (5% efficiency)
GGGCGA (pBS plasmids)	GCTTTG (1% efficiency)
GCCGAA	

Composite 5' primers with the SP6 promoter sequence have also been used successfully. There are also constraints on the 5' sequence of the transcripts, but the optimal sequences (GAATA, GAACA, and GAAGA) are different than those with T7 polymerase *(11–13)*.

8. The 3' end of the transcripts should be exactly defined by the 5' end of the downstream primer. However, template-independent addition of 1–2 nt during transcription usually generates populations of RNAs with different 3' ends. The proportion of each residue at the 3' end may depend on context and is influenced by the relative concentration of each ribonucleotide, limiting ribonucleotide being less frequently incorporated *(2,6,8,14)*.

9. The order of addition of components may be changed. Remember that dilutions of RNasin are very unstable in the absence of DTT, and that DNA should not be added to undiluted 10X TB. The temperature of all components must be at least at 25°C, to avoid precipitation of the DNA:spermidine complex. Since the radioactive nucleotides, which constitute half of the reactions, are provided in well-insulated vials, they can take up to 10–15 min after thawing to reach an acceptable temperature. I routinely incubate all components, except the RNasin and the polymerase, at 30°C for 15–20 min. When more than one probe is to be synthesized, a reaction mixture with all components is added to the DNA template. It is possible to reduce the total volume to 8 µL, in order to conserve materials; however, since the enzymes are very sensitive to surface denaturation, these incubations are done in 400-µL vials.

10. For SP6 polymerase, use 5–10 U/µg of plasmid DNA (size: 3–4 kbp). For T7 and T3 polymerases, use 10–20 U/µg of plasmid DNA.

11. The purification step (**Subheading 3.1.3., step 4**) may not be required, and some investigators use the transcription mixtures directly in hybridization assays. However, low backgrounds, consistency, and quantization of the newly synthesized RNAs probably justify the additional time and effort.

12. More than 50% of the labeled ribonucleotide is routinely incorporated, and often greater than 80%. An input of 50 µCi may therefore yield up to 4×10^7 Cerenkov-cpm of RNA. This represents 100 pmoles of UMP and, assuming no sequence

bias (i.e., 25% of residues are uracil residues), 130 ng of RNA (specific activity: 3×10^8 Cerenkov-cpm/μg).

Similar results are obtained with labeled CTP and GTP, although GTP rapidly loses its incorporation efficiency on storage. ATP is not routinely used because of its higher apparent K_m for SP6 polymerase on linear templates *(2,15)*. UTP is particularly stable and can be used even after several weeks, once radioactive decay is taken into account. With each labeled ribonucleotide, the initial concentration must be > 12.5 μ*M* to ensure efficient incorporation and to prevent polymerase pausing. Transcripts destined to be translated must incorporate a 5' 7mG cap structure. This is usually achieved by performing the transcription in the presence of 500 μ*M* of a cap analogue dinucleotide and 50 mM GTP. Since the cap analogue can also be used during elongation, labeled GTP should not be used to calculate incorporation efficiency.

A variety of other labeled or unlabeled probes may also be made for use in *in situ* hybridization (*see* **refs. 16–20** and Chapter 21).

For the synthesis of large quantities of RNA (more than 10 RNA transcripts per DNA template molecule), each ribonucleotide is added at a final concentration of 0.5–1 m*M*; trace amounts of labeled UTP should be included to calculate transcription efficiency and to verify the size of the transcript. Total volume is increased to reduce the concentration of plasmid DNA to 30–50 μg/mL. Additional modifications of the standard protocol have been described, and include the use of HEPES-KOH pH 7.5 at 120 m*M* (SP6 polymerase), 200 m*M* (T7 polymerase), or 300 m*M* (T3 polymerase) *(21)*. Furthermore, the T4 gene 32 protein can increase transcription efficiency when added at 10 μg/μg of template DNA (D. Caput, personal communication). It has been recently reported that a reduction of premature termination and an increased synthesis of large full length transcripts (size > 1 kb) can be obtained by performing the transcription at 30°C or at room temperature *(4,22)*.

13. Large transcripts that do not enter the polyacrylamide gel are diagnostic of incompletely linearized templates. Small transcripts are indicative of extensive pausing or premature termination. In addition to the natural T7 termination signal, a site located in the coding region of the PTH gene has been extensively characterized *(23,24)*. Mutant T7 RNA polymerases which show reduced termination at natural sites have been constructed by Drs. D. H. Lyakhov and W. T. McAllister (SUNY, Brooklyn, NA).

14. Hybridization of dot-blots is often used to quantitate mRNA levels with large number of samples. To accurately quantitate specific hybrids, it is necessary to include as negative control total RNA from cells that do not express the transcript under examination.

15. The intensity provided by transilluminators, a common source of UV light, varies with time, and excessive crosslinking can severely reduce hybridization efficiency (**Fig. 1A**). Furthermore, UV-crosslinking can increase crosshybridization to rRNAs, possibly by inducing covalent cross-linking with the probe. UV irradiation should thus be limited to membranes subjected to multiple rounds of

hybridization. An apparatus commercially available from Stratagene provides a means to control the dose of UV. The appropriate dose of UV is probably different for different membranes and has a broader range for wet than for dry membranes.

16. The major variables in optimizing signal-to-noise ratio are the temperature of hybridization and the temperature of the stringency washes (**Fig. 1B**). Many probes can be hybridized at 58°C, although the temperature must be increased to 68°C for certain pairs of probe and target RNAs. Stringency washes are usually performed at 70 or 75°C. It is difficult to increase stringency much further, since most waterbaths cannot maintain accurate temperatures at or above 80°C. For the detection of DNA targets, hybridization is usually performed at 42°C and the stringency washes are at 65°C.

17. RNA probes are usually 200–800 nt long. With 600 nt long RNA probes, a hybridization plateau is achieved in 20 h at 58°C with 2.5 ng/mL of probe. This represents an input of 7.5×10^5 Cerenkov-cpm/mL. The probe concentration can be increased to 10 ng/mL, and the hybridization time decreased to 5 h without increasing background.

18. The probes should be 100–400 nt long, and should include at least 10 nt which are not complementary to the target RNA. Residual template DNA generally produces a trace of full length protected probe that must be distinguishable from the fragment protected by the target RNA (**Fig. 2**).

19. It is often useful to decrease the specific activity of the probe: more RNA is synthesized at the resulting higher ribonucleotide concentration, the probes are less susceptible to radiolysis, and less radioactivity is used. The following guideline can be used to alter the specific activity of the probes according to the sensitivity required:

Target RNA abundance	Unlabeled UTP	[^{32}P]-UTP	Probe/sample[a]
High	100 μ*M*	2.5 μ*M*, 10 μCi	6–12 Kcpm
Moderate	10 μ*M*	2.5 μ*M*, 10 μCi	60–120 Kcpm
Low	—	12.5 μ*M*, 50 μCi	300–600 Kcpm

[a]The amount of probe required per sample for the detection of the target (*see* **Note 21**).

20. Purification of is often necessary for maximal sensitivity or for mapping purposes. The transcription reaction can be directly loaded on the preparative gel after DNase digestion of the template, provided that enough EDTA is present in the sample buffer to chelate all the magnesium. Omitting the DNase digestion of the template results in higher amounts of fully protected probe in the assay.

21. The probe must be in excess over the target RNA (*see* **Fig. 2A**). An input of 1–2 ng of a 300-nt probe in a total volume of 30 μL will drive the hybridization of target RNA to completion (4–8 $T_{1/2}$) in approximately 16 h. Shorter hybridizations can be performed but require higher probe input (R_0) to achieve the same extent of saturation, i.e. to maintain the $R_0 \times T_{1/2}$ value.

22. To facilitate the RNase digestion step, each sample should contain the same amount of total RNA. For very low abundance target RNA, the amount of sample

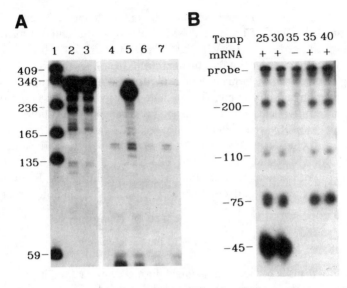

Fig. 2. RNase protection assay. **(A)** Discrimination between target-specific signal and complete probe protection by residual DNA. Detection of PN-I mRNA in total RNA from murine tissues *(30)*. The probe (310 nt long) was gel purified. Lane 1: Size markers. Lane 2: purified probe. Lane 3: probe hybridized and processed without RNase digestion. This part of the gel was autoradiographed for 6 h. Lane 4: control hybridization with 10 µg of tRNA; traces of fully protected probe are visible. Lane 5: 10 µg of RNA from seminal vesicles, an abundant source of PN-I mRNA; the specific protected fragment is 260 nt long. Lane 6: 10 µg of liver RNA, which does not contain detectable levels of PN-I mRNA. Lane 7: 10 µg of testis RNA, which contains trace levels of PN-I mRNA. This part of the gel was autoradiographed for 24 h. **(B)** Effect of hybridization temperature on the detection of short complementary RNAs. The 5' ends of phage T4 gene 32 transcripts in total RNA from bacteria carrying a gene 32 expression cassette were mapped by hybridization to a cRNA probe containing 400 nt of gene 32 upstream sequences *(23)*. The probe was not gel purified, and hybridization were performed at the indicated temperatures. Fully protected probe results from incomplete DNase digestion of the template, and is also visible in the control hybridization without target RNA. The 44 nt protected fragment is no longer detected above 30°C.

RNA may be increased up to 60 µg. Inequalities should be eliminated by addition of tRNA. A negative control sample, containing only tRNA, is always included (**Fig. 2A**, lane 4, **Fig. 2B**, lane 3).

23. The temperature of hybridization must be reduced to detect small or very AU-rich protected fragments. For instance, a 44 nt fragment of a phage T4 gene 32 transcript (containing 35 A/U and 9 C/G) was only protected by performing the hybridization at 25–30°C *(25)* (**Fig. 2B**).

24. Alternatively, the probe and target RNAs may be coprecipitated with ethanol, resuspended in 9 μL of TE, heated for 2 min at 90°C, and chilled on ice. After the addition of 1 μL of 3M NaCl, 0.2M Tris-HCl, pH 7.4, 20 mM EDTA, the hybridization is carried out for 30–60 min at 70°C. Since the probe concentration is higher, there is no formamide and the incubation temperature is higher, the hybridization is driven to completion more rapidly. The RNase digestion can be performed in 50–100 μL (*26*; J. Curran, personal communication).

25. The amount of RNase is determined by the total amount of RNA present in the samples, including that contributed by the probe. I usually add 0.5 μg of pancreatic RNase and/or 0.25 μg of T1 RNase per mg of RNA. In most cases, digestion with pancreatic RNase alone is sufficient. When the probe and the target RNAs are from different species, the extent of homology can be sufficient to generate discrete protected fragments, particularly if digestion is performed with RNase T1 only. The temperature of digestion can be increased to 30–37°C, although this often leads to partial cleavages within the RNA:RNA hybrids.

26. To ensure that the probe remains intact during hybridization, it may be useful to include a parallel control that is hybridized and processed without RNase treatment (**Fig. 2A**, lanes 2 and 3).

27. RNases may also be inactivated by the addition of 330 μL of 4M guanidinium thiocyanate, 25 mM Na-citrate, pH 7.0, and 1M β-mercaptoethanol. Add 20 μg of tRNA, precipitate the RNAs with 660 μL of isopropanol, and centrifuge immediately for 15 min at 13,000×g (*27*; P. A. Menoud, personal communication).

28. Minor shorter protected fragments are often detected, and they may complicate the interpretation of mapping assays *(28)*. To distinguish between digestion artifacts and rare target RNAs which are only partially complementary to the probe, a synthetic sense transcript fully complementary to the probe may be used as a control target RNA *(29)*.

Acknowledgments

I thank P. Vassalli for his early encouragement to use riboprobes for detecting rare mRNAs. Over the last few years, many colleagues, students, and technicians have contributed to the methods outlined in this chapter, including M. Collart, N. Busso, J.-D. Vassalli, H. Krisch, S. Clarkson, J. Huarte, S. Strickland, P. Sappino, M. Pepper, A. Stutz, G. Moreau, D. Caput, M. Prentki, W. Reith, J. Curran, P. A. Menoud, P. Gubler, F. Silva, V. Monney, D. Gay-Ducrest, and N. Sappino. Research was supported by grants from the Swiss National Science Foundation and by the Canton de GenËve. The author is at the University of Geneva Medical School, Department of Pathology, Geneva, Switzerland. E-mail address: dominique.belin@medecine.unige.ch.

References

1. Butler, E. T. and Chamberlin, M. J. (1984). Bacteriophage SP6-specific RNA polymerase. *J. Biol. Chem.* **257,** 5772–5788.

2. Melton, D. A., Krieg, P. A., Rebagliati, M. R., Maniatis, T., Zinn, K., and Green, M. R. (1984). Efficient in vitro synthesis of biologically active RNA and RNA hybridization probes from plasmids containing a bacteriophage SP6 promoter. *Nucl. Acids Res.* **12**, 7035–7056.

3. Davanloo, P., Rosenberg, A. H., Dunn, J. J., and Studier, F. W. (1984). Cloning and expression of the gene for bacteriophage T7 RNA polymerase. *Proc. Natl. Acad. Sci. USA* **81**, 2035–2039.

4. Krieg, P. A. and Melton, D. A. (1987). In vitro RNA synthesis with SP6 RNA polymerase. *Methods Enzymol.* **155**, 397–415.

5. Yisraeli, J. K. and Melton, D. A. (1989). Synthesis of long, capped transcripts in vitro by SP6 and T7 RNA polymerases. *Meth. Enzymol.* **180**, 42–50.

6. Milligan, J. F. and Uhlenbeck, O. C. (1989). Synthesis of small RNAs using T7 RNA polymerase. *Methods Enzymol.* **180**, 51–62.

7. Breaker, R. B., Banerji, A., and Joyce, G. F. (1994). Continuous in vitro evolution of bacteriophage RNA polymerase promoters. *Biochem.* **33**, 11,980–11,986.

8. Milligan, J. F., Groebe, D. R., Witherell, G. W., and Uhlenbeck, O. C. (1987). Oligoribonucleotide synthesis using T7 RNA polymerase and synthetic DNA templates. *Nucl. Acids Res.* **15**, 8783–8798.

9. Roitsch, T. and Lehle, L. (1989). Requirements for efficient in vitro transcription and translation : a study using yeast invertase as a probe. *Biochim. Biophys. Acta* **1009**, 19–26.

10. Schenbon, E. T. and Mierendorf, R. C. (1985). A novel transcription property of SP6 and T7 RNA polymerases:dependence on template structure. *Nucl. Acids Res.* **13**, 6223–6234.

11. Nam, S. C. and Kang, C. (1988). Transcription initiation site selection and abortive initiation cycling of phage SP6 RNA polymerase. *J. Biol. Chem.* **263**,18,123–18,127.

12. Solazzo, M., Spinelli, L., and Cesareni, G. (1987). SP6 RNA polymerase: sequence requirements downstream from the transcription start site. *Focus* **10**, 11–12.

13. Stump, W. T. and Hall, K. B. (1993). SP6 RNA polymerase efficiently synthesizes RNA from short double-stranded DNA templates. *Nucl. Acids Res.* **21**, 5480–5484.

14. Moreau, G. (1991). RNA binding properties of the *Xenopus* LA proteins. Ph.D. thesis, University of Geneva.

15. Taylor, D. R., and Mathews, M. B. (1993). Transcription by SP6 RNA polymerase exhibits an ATP dependence that is influenced by promoter topology. *Nucl. Acids Res.* **21**, 1927–1933.

16. Sappino, A.-P., Huarte, J., Belin, D., and Vassalli, J.-D. (1989). Plasminogen activators in tissue remodeling and invasion : mRNA localization in mouse ovaries and implanting embryos. *J. Cell Biol.* **109**, 2471–2479.

17. Jostarndt, K., Puntschart, A., Hoppeler, H., and Billeter, R. (1994). The use of [33P]-labeled riboprobes for in situ hybridizations: localization of myosin light chain mRNAs in adult human skeletal muscle. *Histochem. J.* **26**, 32–40.

18. Dörries, U., Bartsch, U., Nolte, C., Roth, J., and Schachner, M. (1993). Adaptation of a non-radioactive *in situ* hybridization method to electron microscopy: detection of tenascin mRNA in mouse cerebellum with digoxigenin-labeled probes and gold-labeled antibodies. *Histochem.* **99**, 251–262.

19. Kriegsmann, J., Keyszer, G., Geiler, T., Gay, R. E., and Gay, S. (1994). A new double labeling technique for combined *in situ* hybridization and immunohistochemical analysis. *Lab. Invest.* **71**, 911–917.

20. Egger, D., Troxler, M., and Bienz, K. (1994). Light and electron microscopic in situ hybridization: non-radioactive labeling and detection, double hybridization, and combined hybridization-immunocytochemistry. *J. Histochem. Cytochem.* **42**, 815–822.

21. Pokrovskaya, I. D. and Gurevich, V. V. (1994). In vitro transcription: preparative RNA yields in analytical scale reactions. *Anal. Biochem.* **220**, 420–423.

22. Krieg, P. A. (1991). Improved synthesis of full length RNA probe at reduced incubation temperatures. *Nucl. Acids Res.* **18**, 6463.

23. Belin, D., Mudd, E. A.,Prentki, P., Yi-Yi, Y., and Krisch, H. M. (1987). Sense and antisense transcription of bacteriophage T4 gene 32. *J. Mol. Biol.* **194**, 231–243.

24. Mead, D. A., Szesna-Skorupa, E., and Kemper, B. (1986). Single-stranded DNA blue T7 promoter plasmids. *Prot. Eng.* **1**, 67–74.

25. Macdonald, L. E., Durbin, R. K., and McAllister, W. T. (1994). Characterisation of two types of termination signals for bacteriophage T7 RNA polymerase. *J.Mol.Biol.* **238**, 145–158.

26. Curran, J., Marq, J. B., and Kolakofsky, D. (1992). The Sendai virus nonstructural C proteins specifically inhibit viral mRNA synthesis. *Virology* **189**, 647–656.

27. Hod, Y.(1992). A simplified ribonuclease protection assay. *BioTechniques* **13**, 852,853.

28. Lau, E. T., Kong, R. Y. C., and Cheah, K. S. E. (1993). A critical assessment of the RNase protection assay as a means of determining exon sizes. *Anal. Biochem.* **209**, 360–366.

29. Belin, D., Wohlwend, A., Schleuning, W.-D., Kruithof, E. K. O., and Vassalli, J.-D. (1989). Facultative polypetide translocation allows a single mRNA to encode the secreted and cytosolic forms of plasminogen activators inhibitor 2. *EMBO J.* **8**, 3287–3294.

30. Vassalli, J.-D., Huarte, J., Bosco, D., Sappino, A.-P., Sappino, N., Velardi, A., Wohlwend, A., Erno, H., Monard, D., and Belin, D. (1993). Protease-nexin I as an androgen-dependent secretory product of the murine seminal vesicle. *EMBO J.* 1871–1878.

17

Analysis of RNA by Northern Blotting Using Riboprobes

Rai Ajit K. Srivastava

1. Introduction

To study the expression of a gene in mammalian tissues, it is important to determine the levels of the corresponding mRNA. Several methods have been described for measuring the level of expression of a gene in a tissue. These methods are: *in situ* hybridization using cDNA probe or riboprobe *(1)*, slot-blot hybridization on total RNA isolated from tissues *(2)*, and Northern blotting hybridization using either a cDNA *(3)* or a riboprobe *(4)*. Absolute levels of a specific mRNA are also determined by RNase protection assay using a cDNA probe *(5)* or a riboprobe *(6)*. *In situ* hybridization provides relative expression of a gene, while the slot blot technique is not sensitive enough for accurate measurements of mRNA. The most widely used technique to measure mRNA levels in mammalian tissues is still Northern blotting analysis, in which total RNA or poly(A⁺) RNA are separated by electrophoresis in an agarose gel containing formaldehyde and transferred onto nitrocellulose or nylon membrane. The transferred RNAs on the membrane are then denatured and probed with a labeled cDNA probe or a riboprobe. While a cDNA probe gives the same information as the riboprobe, the sensitivity of the detection of mRNA is increased several fold by using a riboprobe (**ref. *4*** and **Fig. 1**) because of the increased affinity of riboprobe for the complementary sense strand of mRNA and the higher stability of the double-stranded RNA after hybridization. However, the sensitivity of various membranes may marginally differ while using a riboprobe *(4)*. Thus, the higher sensitivity of the riboprobes in Northern blotting analysis makes them a better choice over cDNA probes, especially when detecting a low abundance message. However, care must be taken while using riboprobes for Northern blotting analysis since improper use

From: *Methods in Molecular Biology, Vol. 86: RNA Isolation and Characterization Protocols*
Edited by: R. Rapley and D. L. Manning © Humana Press Inc., Totowa, NJ

Probed with cDNA   Probed withRiboprobe

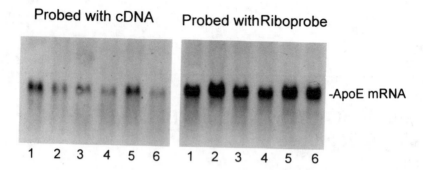

Fig. 1. Northern blot analysis using riboprobe. Six RNA samples (15 μg of each) from mouse liver were electrophoresed on the same gel in two replicates, and separated by cutting the membrane from the middle after transfer for hybridization with mouse apoAI cDNA probe (left panel) or with mouse apoAI riboprobe (right panel). For hybridization with riboprobe 1% nonfat dry milk was used to block the membrane.

of riboprobes may result in very high background, often by irreversible binding of the riboprobe to the membrane. Therefore, the steps shown in this chapter should be followed strictly in order to avoid background noise.

2. Materials

2.1. Preparation of Cellular RNA

1. For RNA analysis all glassware to be used should be properly treated with 0.05% diethylaminoethyl pyrocarbonate in a hood overnight, followed by autoclaving.
2. Deionized water used for preparing solutions or for use in any reaction should be similarly treated.
3. Pipetman used for RNase should be avoided while working with RNA samples.
4. Animal tissue samples or cultured cell line: The animal tissues and cultured cells to be used for RNA preparation should be fresh or quick-frozen in liquid nitrogen and stored at –70°C.
5. Solution A: 4M guanidinium isothiocyanate, 25M sodium acetate, pH 7.0, 0.1M β-mercaptoethanol. Make this solution fresh from the stock solutions every time before use.
6. Solution B: equal volume of 5% sarcosyl and 2M sodium acetate, pH 4.0. Make fresh every time.
7. Salt saturated phenol: Prepare 1M salt-saturated phenol and store at 4°C in a brown bottle to protect from light.
8. Chloroform/isoamylalcohol: Mix chloroform and isoamylalcohol in 49:1 ratio just before use.
9. Isopropanol: molecular biology-grade isopropanol.
10. 70% ethanol: Prepare 70% ethanol using DEPC-treated water and always keep it chilled at –20°C freezer.
11. SDS solution: Prepare 0.2 % SDS solution and store at room temperature.

2.2. Preparation of Riboprobe

1. Template plasmid: The cDNA fragment to be used should be subcloned in the polylinker region of a vector. The polylinker region should be flanked by T7 and SP6 RNA polymerase promoters. The recombinant plasmid should be linearized with appropriate restriction enzyme and purified. (*See* **Note 1** and **Fig. 2.**)
2. 10X transcription buffer: 400 mM Tris-HCl pH 7.5, 60 mM MgCl$_2$, 100 mM NaCl, and 20 mM spermidine. Store in aliquots of 100 µL at –20°C.
3. Nucleotide solution: 10 mM solution of each NTP (can be purchased, *see* **Note 2**). Store at –20°C.
4. Nucleotide mix: Prepare nucleotide mix as follows: A 1:1:1 ratio mix of ATP, GTP, and UTP. Store at –20°C.
5. TCA solution: 10% trichloroacetic acid, 1.5% sodium pyrophosphate. Store at 4°C.

2.3. Agarose Gel Electrophoresis

1. Agarose: Use molecular biology-grade agarose.
2. Deionized formamide: Take 20 mL formamide in polypropylene, add 4 g mixed-bed resin and stir for at least 2 h. Aliquot and store at –20°C.
3. 10X MOPS: 0.2M MOPS, 0.05M sodium acetate, 0.01M EDTA. Adjust pH to 7.0 with 1M NaOH. Filter sterilize and store in a refrigerator in the dark.
4. Formaldehyde: Use commercially supplied formaldehyde, which is 37% (v/v) formaldehyde.
5. Ethidium bromide. Prepare 0.5 mg/mL ethidium bromide in DEPC-treated water. Store refrigerated in the dark.
6. Gel loading dye: bromophenol blue 0.001%, xylene cyanol 0.001% in DEPC-treated water.
7. 20X SSC: 3M NaCl, 0.3M Sodium citrate. Adjust pH to 7.0 with HCl. Sterilize by autoclaving.
8. RNA sample buffer: 50% deionized formamide, 6% formaldehyde, 1X MOPS.
9. Membrane: Use nitrocellulose or nylon membrane.

2.4. Hybridization

1. Prehybridization and hybridization buffer: 100 mM Tris-HCl, pH 7.4, 10 mM EDTA, 600 mM NaCl, 1.5% nonfat dry milk (purchased from a supermarket), and 50% formamide. Filter through 0.45 µm nitrocellulose filter and incubate overnight at 70°C (*see* **Note 3**). Then add 10% SDS to bring final concentration of SDS to 1%.

3. Methods
3.1. Isolation of Cellular RNA

1. Homogenize tissue samples (0.2 g) with 1 mL of solution A. When using cultured cells, wash cells with PBS (phosphate-buffered saline) and add 1 mL of solution A for each 3–5 × 10^6 cells. With the help of a 1 mL Pipetman mix cells with solution A thoroughly by pipeting in and out several times.
2. To the homogenized tissue samples or the cultured cells, add 300 µL of solution B and 1 mL salt-saturated phenol followed by vigorous vortexing for 15 s.

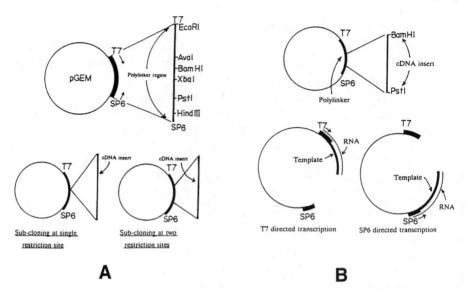

Fig. 2. (A) Subcloning of the cDNA fragment into the polylinker region of a plasmid vector. Shown are subcloning either into a single restriction enzyme site or two different sites. (B) Linearization of recombinant plasmid for in vitro transcription to synthesize riboprobe.

3. Add 250 µL of chloroform:isoamyl alcohol (49:1) and vortex vigorously for 15 s.
4. Allow the mixture to sit in an ice bath for 5 min. and centrifuge at 12,000g for 15 min at 4°C.
5. Remove upper aqueous layer carefully to another tube avoiding turbid materials from the interface (see Note 4).
6. Re-extract the supernatants with equal vol of phenol/chloroform, and transfer the upper layer to another tube.
7. Add equal vol of isopropanol and leave on ice bath for 30 min or at –20°C for 15 min and pellet RNA by centrifugation at 12,000g for 15 min (see Note 5).
8. Wash RNA precipitate with 500 µL of chilled 70% ethanol.
9. Dry RNA for 5 min in a Speedvac.
10. Dissolve dried RNA samples in 0.2% SDS, determine the A_{260}/A_{280} of appropriately diluted RNA sample, which should usually be more than 1.8 (see Note 6 and Chapter 12). After determining the concentration of RNA, store in aliquots at –70°C (see Note 7).

3.2. In Vitro Synthesis of Riboprobe

1. For the in vitro synthesis of riboprobe, usually 0.5 to 1 µg of the linearized and purified template recombinant plasmid is used (see Note 8). To a DEPC-treated 1.5 mL Eppendorf tube, add the following ingredients at room temperature in the following order: 3 µL of 10X transcription buffer, 3 µL of 0.1M DTT, 1 µL of RNasin (ribonuclease inhibitor), 3 µL of nucleotide mix, 5 µL of linearized plasmid (0.5–1 µg),

2.5 μL of α-^{32}P[CTP] (Amersham, 20 mCi/mL), sterile water to make the total volume to 30 μL and 1 μL of T7, SP6 or T3 RNA polymerase (*see* **Note 9**). Mix the contents by brief centrifugation on a table-top microcentrifuge. Incubate at 37°C for 1 h.

2. Stop the transcription reaction by the addition of 1 μL of RNase-free DNase I (RQ1 from Promega or other RNase-free DNase I works fine). Incubate at 37°C for an additional 15 min.
3. Add 1 μL of RNasin and mix by brief vortexing.
4. Remove unincorporated radionucleotides, by Sephadex G-50 column as described in **Subheading 3.2.6**. Both the DNA as well as RNA columns from BMB or from other companies work fine (*see* **Note 10**). Drain the buffer off from the Sephadex column by gravity in a clean tube supplied with the column.
5. Centrifuge the column at 200*g* for 2 min in a swinging bucket rotor to drain off the remaining buffer of the column. Discard the tube in which buffer was collected.
6. Place the Sephadex column in another clean tube supplied with the column and carefully apply the transcribed riboprobe to the center of the Sephadex beads (*see* **Note 11**).
7. Centrifuge again at 200*g* for 2 min. Riboprobe collects in the clean tube.
8. Determine the quality of the riboprobe (*see* **Note 12** and **Fig. 3**).
9. Use riboprobe fresh or keep stored in aliquots at –70°C (*see* **Note 13**).

3.3. Electrophoresis and Transfer of RNA

1. For a 11 × 14 cm gel prepare agarose in a 250 mL Erlenmeyer flask by melting 1.2 g of agarose in 73 mL of DEPC-treated water in a microwave.
2. Cool the contents to 50–55°C and add 10 mL of prewarmed (37°C) 10X MOPS buffer and 16.2 mL of 37% formaldehyde.
3. Mix gently but thoroughly, and pour into the gel rig, which has tape on both ends and a comb.
4. Remove the bubbles, if any, with a sterile Pasteur pipet.
5. Let the gel solidify for at least 20 min.
6. Take 15 μg of total RNA in an Eppendorf tube and add 15 μL of RNA sample buffer. Incubate at 60°C for 15 min.
7. Cool on ice for 5 min add 2 μL of gel loading dye and mix.
8. Centrifuge briefly in a microcentrifuge to collect samples at the bottom of the tube.
9. Prepare 1X MOPS in DEPC-treated water, and prerun the gel for 20 min at 80 V.
10. Load RNA sample prepared as above in the gel slots, and run at constant voltage of 90 for 3–4 h or until the tracking blue dye reached 1 in. from the bottom of the gel.
11. After electrophoresis, stain the gel either with ethidium bromide or with acridine orange to visualize 28S and 18S ribosomal RNA (*see* **Note 14**).
12. Cut a piece of Gene Screen membrane of the approximate size of the gel, and soak in 10X SSC buffer.
13. Set-up the transfer using the standard method of capillary transfer. Soak a Whatman paper and put it on the plate so that both ends of the Whatman paper are sitting in the buffer.
14. Flip the gel over so that the loading side faces downward (*see* **Note 15**) and place on the Whatman paper wick. Make a nick on the one corner of the gel to mark

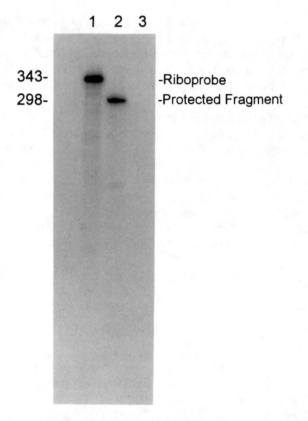

Fig. 3. Determination of quality of the riboprobe. A 298-bp mouse apoE riboprobe was subcloned into the *Pst*I site of the pGEM3Zf(+) vector, and linearized with *Hind*III to prepare riboprobe. Lane 1, riboprobe lane 2, riboprobe hybridized with 5 μg of total mouse liver RNA lane 3, riboprobe hybridized with 5 μg of yeast tRNA.

    lane 1. Place the membrane on top of the gel, and remove all air bubbles with the help of a 10 mL pipet by rolling from one side to another.

15. Put Whatman paper on the top of the membrane followed by a stack of paper towel (usually 3-in height).

16. Place a small glass plate on top of the paper towel followed by a 200–300 g weight.

17. Let the transfer proceed overnight.

18. Next morning take out the membrane and rinse with 2X SSC. Check the gel and the membrane under UV light for complete transfer of the RNA.

19. Take a photograph of the membrane and mark 28S and 18S ribosomal RNA on the membrane with a pencil as shown in **Fig. 4**.

20. Bake the membrane at 80°C for 2 h in a vacuum-oven, and put in a plastic sandwich bag.

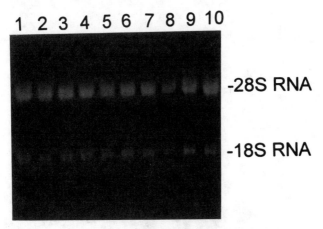

Fig. 4. Ethidium bromide stained gel of total RNA prepared by using the method described in this section. Fifteen micrograms of ten different RNA samples were prepared and electrophoresed. The positions of 28S and 18S RNA are marked.

3.4. Hybridization

1. For a 11 × 13 cm membrane take 12 mL of hybridization buffer and mix with denatured salmon sperm DNA (150 µg/mL) (*see* **Note 16**).
2. Pour the prehybridization buffer in the bag containing the membrane, and seal the bag avoiding any entrapment of air bubbles (*see* **Note 17** and **Figs. 1** and **5**).
3. Place the blot in a shaking water bath at 60°C for 3 h.
4. After the prehybridization is complete, heat the riboprobe (3–5 × 10⁶ cpm) together with 100 µL of salmon sperm DNA (5 mg/mL) at 90°C for 5 min and cool on ice bath for 5 min.
5. Add denatured riboprobe in the hybridization bag and seal the bag avoiding any entrapment of bubbles.
6. Allow hybridization to proceed overnight at 60°C.
7. On the next day, cut open the bag and pour the radioactive contents of the bag in a radioactive waste disposal container. Add 20 mL of 1X SSC/0.2% SDS in the hybridization bag and after rinsing the membrane discard the solution in the radioactive waste disposal container.
8. Wash with 200 mL of 1X SSC/0.2% SDS at room temperature for 10 min, repeat twice.
9. Wash with 200 mL of 0.1X SSC/0.1% SDS for 30 min at 65°C twice.
10. Rinse with 100 mL of 0.1X SSC and let it dry at room temperature. Do not dry completely.
11. Cover the membrane with Saran wrap and expose to X-ray film.

4. Notes

1. Several plasmid vectors are currently available that have two different RNA polymerase promoters flanking the polylinker region in which the desired cDNA fragment is subcloned for the purpose of synthesizing RNA tran-

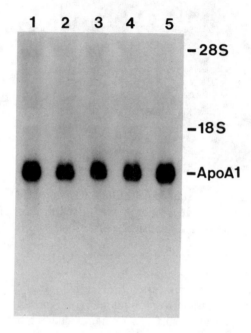

Fig. 5. Northern blot analysis using riboprobe. Mouse RNA were electrophoresed as described in caption to **Fig. 1** and probed with mouse apoAI riboprobe using 1.5% nonfat dry milk for blocking the membrane.

scripts. Some vectors like pGEM3Z series have T7 and SP6 RNA polymerase promoters, and some vectors like BlueScript have T7 and T3 RNA polymerase promoters. The orientation of the cDNA insert in a vector should be determined by sequencing the double-stranded plasmid *(7)* in order to determine where to linearize the recombinant plasmid for riboprobe synthesis. An illustration has been provided in **Fig. 2**.

2. Nucleotide solutions can be purchsed from Promega, Ambion,or BMB. All of these nucleotide solutions have a pH of 7.5.

3. Alternatively, the prehybridization solution can be treated for 1 h at 90°C.

4. While transferring the upper layer containing RNA, care must be taken to avoid taking any material from the interface because of the presence of proteins that also contain ribonucleases. To be safe, one should carefully take only 2/3 of the upper layer slowly and without disturbing the interface.

5. Precipitation of RNA with isopropanol at 4°C for overnight gives better yield because of the complete precipitation of all sizes of RNA.

6. Instead of 0.2% SDS, the RNA can be dissolved in DEPC-treated deionized water. If the $A_{260/280}$ ratios of RNA samples are lower than 1.8, it suggests that the RNA

are contaminated with proteins or ribonucleoprotein complexes *(8)*. It has been noticed that the method described here does not require proteinase K digestion for separating RNA from the ribonucleoprotein complexes. The quality of the RNA from a typical isolation is shown in **Fig. 4.**

7. The quality of the RNA prepared can be checked by performing agarose gel electrophoresis as shown in **Fig. 2.** The ratios of 28S RNA to 18S RNA should be usually 2:1. When degradation of RNA occurs, the ratio decreases (*see* Chapter 12).

8. After linearization of the recombinant plasmid by an appropriate restriction endonuclease, it should be purified by phenol-chloroform extractions, ethanol precipitation, and dissolved in DEPC-treated water.

9. Check the recombinant vector for appropriate RNA polymerase promoter flanking the polylinker region.

10. BMB provides Sephadex G-25 and G-50 columns exclusively for RNA, although DNA columns also works well. Sephadex columns from other companies also work equally as well.

11. After draining the buffer from the Sephadex column by centrifugation, the gel beads become one solid aggregate. The riboprobe reaction mixture should be applied to the center of the aggregate by avoiding any contact with the column sides. This will allow all the reaction mixture to go through the beads at the time of centrifugation, and will yield a pure riboprobe.

12. The quality of the riboprobe can be determined by running a sequencing gel and observing an appropriate size singe band of riboprobe, and after hybridization with the total RNA it should protect an appropriate size RNA as shown in **Fig. 3.** Alternatively, riboprobes can be precipitated with 10% TCA and filtered through glass fiber filter and the filter counted. There should not be more than 10% difference between the initial riboprobe count and after TCA precipitation.

13. After synthesizing the riboprobe it can be stored frozen at −70°C for up to 2 wk without any loss in the quality of the riboprobe.

14. It is always desirable to stain gel with acridine orange and then destain to visualize 28S and 18S RNA, since acridine orange does not interfere with the transfer of RNA. If the gel is stained with ethidium bromide, proper destaining avoids any problem of RNA transfer.

15. It has been consistently observed that the transfer of RNA is always complete when flipping over the gel. This is particularly important if the gel has been stained with ethidium bromide.

16. Denature salmon sperm DNA by boiling for 10 min. A stock of salmon sperm DNA can be prepared and stored in aliquots at −20°C.

17. One percent nonfat dry milk has been used by others for Northern blotting analysis using riboprobe. However, using 1.5% nonfat dry milk gives less backgound as compared to 1% nonfat dry milk. The Northern blot analysis shown in **Fig. 1** was performed with 1% nonfat dry milk, and that shown in **Fig. 5** with 1.5% nonfat dry milk. It is seen that blocking the membrane with 1.5% nonfat dry milk gives lower background.

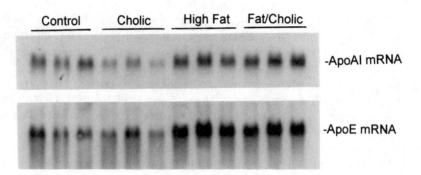

Fig. 6. Reprobing a blot with a riboprobe. The blot shown here was first probed with apoE riboprobe to detect apoE mRNA (1200 nt), stripped off (stripping does not completely remove apoE riboprobe), and then probed with apoAI riboprobe to detect apoAI mRNA (1000 nt). Groups of mice were fed the diets indicated in the figure, hepatic RNA isolated and analyzed by Northern blotting.

18. Unlike with the use of the cDNA probe, use of riboprobes in the Northern blotting analysis makes it difficult to reprobe the membrane for the mRNA of similar sizes, since riboprobes are not stripped off completely from the membrane even after boiling the membrane, especially when the concentrations of the target mRNA are very high (4). However, different sizes of mRNA can be reprobed on the same blot after stripping-off as shown in **Fig. 6.**

References

1. Polacek, D., Becman, M. W., and Schrieber, J. R. (1992) Rat ovarian apolipoprotein E: localization and gonadotropin control of messenger RNA. *Biol. Reprod.* **46**, 65–72.
2. Davis, L. G., Dibner, M. D., and Battey, J. F. in *Basic Methods in Molecular Biology*, Elsvier, New York, pp. 147–149.
3. Twoney, T. A. and Krawetz, S. A. (1990) Parameters affecting hybridization of nucleic acids blotted onto nitrocellulose membrane. *BioTechniques* **8**, 478–481.
4. Srivastava, R. A. K. and Schonfeld, G. (1991) Using riboprobes for Northern blotting analysis. *BioTechniques* **11**, 584–587.
5. Williams, D. L., Newman, T. C., Shelnes, G. S., and Gordon, D. A. (1986) Measurements of apolipoprotein E mRNA by DNA-excess solution hybridization with single-stranded probes. *Methods Enzymol.* **128**, 671–689.
6. Srivastava, R. A. K., Pfleger, B., and Schonfeld, G. (1991) Expression of low density lipoprotein receptor, apolipoprotein AI and apolipoprotein AIV mRNA in various mouse organs as determined by a novel RNA-excess solution hybridization assay. *Biochim. Biophys. Acta* **1090**, 95–101.
7. Mierendorf, R. C. and Pfefferl D. C. (1987) Direct sequencing of denatured plasmid DNA. *Methods Enzymol.* **152**, 556–562.
8. Srivastava, R. A. K., Srivastava, N., and Schonfeld, G. (1992) Expression of low density lipoprotein receptor, apolipoprotein AI, AII, and AIV in various rat organs utilizing an efficient and rapid method for RNA isolation. *Biochem. Intern.* **27**, 85–95.

18

RNA Quantitative Analysis
from Fixed and Paraffin-Embedded Tissues

Giorgio Stanta, Serena Bonin, and Renè Utrera

1. Introduction

As already reported in a previous chapter of this book, it is possible to extract and analyze RNA from fixed- and paraffin-embedded tissues (*see* Chapter 5). For many purposes, a simple qualitative analysis of the presence of a specific RNA in tissues may be sufficient, for instance, to establish the persistence in tissues of the genome of an RNA virus. However, when gene expression levels are needed, quantiation of the specific mRNA is an important task.

The major problem with paraffin-embedded tissues is the extent of degradation of the extracted RNA. Usually in routinely treated formalin-fixed material the available fragments of RNA range between 100 and 200 bases (*1*), although the level of degradation can vary in different tissues or in different samples of the same tissue, even when they have been processed in the same laboratory. This means that for any analysis, the amenable sequences are short and therefore it is recommended to study sequences of 100 bases or less.

To obtain quantiation of the RNA standardization of the level of degradation of the RNA for each sample is needed by comparison with a different internal mRNA. Other types of effective quantization such as competitive analysis cannot be performed with degraded RNAs because of the different levels of degradation.

To perform relative quantiation of a specific mRNA in paraffin-embedded tissue, the sequence is transcribed to cDNA, and then a short fragment of maximum 100 bases in length is amplified. The amplification must be performed in conditions in which there is a linear relationship between the log of the initial quantity of RNA and the log of the quantity of the amplified

From: *Methods in Molecular Biology, Vol. 86: RNA Isolation and Characterization Protocols*
Edited by: R. Rapley and D. L. Manning © Humana Press Inc., Totowa, NJ

product. The quantitative results are standardized with a reference sequence, we use the β-actin or GAPDH mRNA sequences. This comparison must be performed in every sample to normalize the level of degradation of the RNA.

2. Materials

1. AMV Reverse transcriptase.
2. AMV buffer 1X: 50 mM Tris-HCl, pH 8.3 (42°C), 50 mM KCl, 10 mM MgCl$_2$, 10 mM DTT, 0.5 mM spermidine. The buffer is stored in aliquots at –20°C as 5X stock solution.
3. The dNTPs are aliquoted together in a stock solution 10 mM each and stored at –20°C.
4. Oligonucleotides: Sense and antisense primers for PCR amplification are aliquoted at –20°C in stock solutions 30 pmol/μL. The oligonucleotide used as probe is aliquoted at –20°C at the concentration of 250 ng/μL.
4. Taq Polymerase: may be obtained from Perkin-Elmer.
5. PCR buffer 1X: 10 mM Tris-HCl, pH 8.3, 55 mM KCl. The buffer is stored at –20°C as 10X stock solution.
6. SSC (20X): 3M NaCl, 0.3M tri-sodium-citrate. Adjust pH to 7.0 with 1M HCl. Store at room temperature.
7. Dye for dot-blot: 0.25% bromophenol blue, 2.5% ficoll in water. Store at 4°C.
8. SSPE (20X): 3M NaCl, 0.2M NaH$_2$PO$_4$, H$_2$O, 0.02M EDTA. Adjust pH to 7.4 with 10M NaOH. All the components are solubilized by the autoclaving. Store at room temperature.
9. Denhart's Solution (50X): 5 g polyvinylpyrrolidone, 5 g BSA, 5 g ficoll 400, 250 mL of sterile water. Store at –20°C.
10. T4 polynucleotide kinase.
11. T4 polynucleotide kinase buffer (1X): 70 mM Tris-HCl, 10 mM MgCl$_2$, 5 mM DTT, pH 7.6. Stock solution 10X stored at –20°C.
12. Sephadex G25: Add 30 g of Sephadex G25 to 250 mL of sterile water in a bottle. Make sure that the powder is well dispersed. Let stand overnight at room temperature. Decant the supernatant water and replace with an equal volume of water. Store at 4°C.
13. Nylon Membrane Hybond N$^+$ (Amersham).

3. Methods

To quantify an mRNA: Three oligonucleotides are synthesized, two in mRNA sense and one in antisense orientation, which is used for the reverse transcription. The first sense and the antisense oligonucleotide, used to amplify the sequence, must be in two successive exons of the gene in order to distinguish it from amplification of contaminant genomic DNA. The second sense oligonucleotide spans over the two exons junctions and is used as a probe for the amplified product. The segments of mRNA studied are usually very short, between 75 and 100 bases, because of the degradation of the RNA. The amplification conditions must be determined very carefully because a relative quantization is only possible when the log of total RNA (in ng/100 µL) has a linear relationship with the log of the counts (cpm) of the specific amplification product. These conditions depend on the amount of RNA and the number of amplification cycles. Moreover, the RNA degradation must be normalized using the quantitative analysis of a well-expressed mRNA since the level of degradation varies among samples.

3.1. RNA Extraction

See Chapter 5, on RNA extraction from fixed and paraffin-embedded tissues.

3.2. Reverse Transcription

After spectrophotometic quantiation (*see* **Note 1**), the total RNA of different samples are diluted to the same concentration, to use similar quantities of target RNA for each sample.

1. Treat RNAs with 2.5 U of AMV reverse transcriptase (Promega) in 10 µL final volume containing AMV buffer 1X, 1 m*M* dNTPs and 15 pmol (0.5 µL) of downstream antisense primer.
2. Incubate reaction at 42°C for 60 min.

3.3. Amplification

1. Add 40 µL of master mix containing 1X PCR buffer without Mg, 15 pmol of upstream primer, and 1.2 U of AmpliTaq Polymerase.
2. Denature for 3 min at 95°C.
3. Carry out cycling for 5 cycles of 95°C/1 min, 55°C/1 min, 72°C/1 min and between 30 and 55 cycles of 95°C/30 s, 55°C/30 s, 72°C/30 s in a Thermocycler bearing a heated cover system to avoid aerosol.

3.4. Dot-Blot (see *Note 2*)

1. Equilibrate the Hybond N$^+$ membrane (Amersham) in SSC 10X for 10 min.
2. Denature 20 µL of amplified material for 10 min at 95°C. Chill on ice and add to each sample 30 µL of SSC 20X and 1 µL of dye for dot-blot.
3. Spot the samples on membrane using a dot-blot apparatus (*see* Chapter 14).
4. Air-dry the membrane and crosslink twice the DNA with a UV-Stratalinker (Stratagene, La Jolla, CA) for 120 mJ at the time.

3.5. Hybridization and Counting

For the hybridization an oligonucleotide internal to the amplified sequence, 5' end labeled with ^{32}P isotope is used as probe.

1. The labeling is performed in kinase buffer using 500 ng of the oligonucleotide at 37°C for 60 min.
2. Purify the labeled probe using a G-25 Sephadex minicolumn.
3. Prepare the Sephadex G25 column using as column a Costar centrifuge tube filter (Costar, Cambrige, MA): Load the small column with the Sephadex G25 prepared as described (*see* Chapter 26).

4. With a vacuum water pump, carefully package the column—at the end the gel must be uniform without any crack.

5. Add to the column 100 µL of water and spin down in a centrifuge for 15 s at 14,000g. Repeat this step until the volume left in the bottom of the tube is again 100 µL. At this point the column is equilibrated and you can load it with 100 µL of the radiolabeled probe.

6. Centrifuge at exactly the same speed and for exactly the same time as before collecting the effluent from the tube. The unincorporated [γ^{32}P]dATP remains in the column (*2*).

7. Perform hybridization over night in 5X SSPE, 1X Denhart's solution, and 0.5% SDS at 50–54°C for most of the probes of 24–30 nucleotides with an AT:GC ratio close to 1.

8. Perform two washings at room temperature with SSPE 2X, SDS 0.1% each for 10 min, one at 60°C with SSPE 1X, SDS 0.1% and one at 60°C with SSPE 0.1X and SDS 0.01% each for 15 min. The temperatures of hybridization and washing can be changed according with the T_m of the probe.

9. Autoradiography of the membrane is suggested before cutting it to see if the spots are clearly defined and to check the absence of a specific highly radioactive spots on the membrane owing to the use of unwashed gloves during manipulation or to other dirt on the membrane.

10. Cut every spot with the same area of membrane and count in a single vial using a β-scintillation counter. Count also the same area of membrane without spotted samples (empty membrane).

3.6. Search for Linearity Conditions

There is a linear relationship between the log of the quantity of total RNA and the log of the amplified product. We checked these conditions using an in vitro transcribed T7 RNA of 456 bases from retinoblastoma mRNA sequence. Since the linearity is also obtained with the paraffin-embedded tissue RNA (**Fig. 1**), we conclude that no inhibition is present using our extraction methodology.

The linearity conditions for RNA extracted from formalin-fixed and paraffin-embedded tissue must be calculated taking into account the fact that most of the RNA is highly degraded. First of all, a constant quantity of total RNA from paraffin-embedded tissue (about 400 ng in most of the cases) must be amplified for a specific sequence, after reverse transcription, with a different number of cycles. The linearity between the log of RNA and the number of cycles is usually conserved in RNA from paraffin-embedded tissues even at higher number of cycles (up to 70), while this is not the case for RNA from fresh tissues in which the linearity is lost at a lower number of cycles. Usually in paraffin-embedded tissues the linearity is present in most mRNA sequences between 30 and 60 cycles of amplification (**Fig. 2**). We often use an amplification of 40 cycles for the more abundant and 55 cycles for less abundant mRNAs.

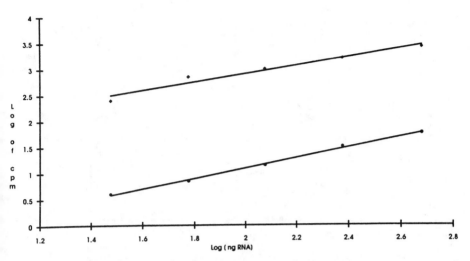

Fig. 1. The lower line corresponds to amplification for 25 cycles of a sequence of 75 bases of mRNA of the Rb gene, after reverse transcription, from an in vitro transcribed RNA of 456 bases, using 1–16 pg of the RNA. The upper line represents the same type of amplification using increasing amounts of total RNA (30–500 ng) extracted from paraffin-embedded tissues to which 4 pg of the in vitro transcribed Rb RNA were added to each sample. The two lines behave almost in a parallel way, showing that no inhibition because of the extraction procedure is present.

The second step to determine the linearity conditions consists in the repetition of the amplification with the chosen number of cycles and with increasing quantities of RNA. We normally use: 7.8, 15.6, 31.25, 62.5, 125, 250, 500, 1000, and 2000 ng of total RNA for 100 µL of amplification solution. The logarithms of cpm of the hybridized amplified product show a linear relationship for most mRNAs in the range between the log of 125 and 1000 ng of total RNA. For mRNAs with very low or very high expression levels, the number of cycles and quantity suggested must be sometimes changed and adapted. This linearity conditions analysis must be performed for each new sequence studied. When the right conditions of amplification (number of cycles and initial quantity of RNA) are established, it is possible to analyze a large case study of tissues in the same conditions. The quantitation can be performed with membrane hybridization as reported here or with the capillary electrophoresis quantization *(3)*.

3.7. Standardization of RNA Degradation Level

The comparison among samples cannot be made directly, even if the initial quantity of RNA and the tissue type are the same. This is because of the possible

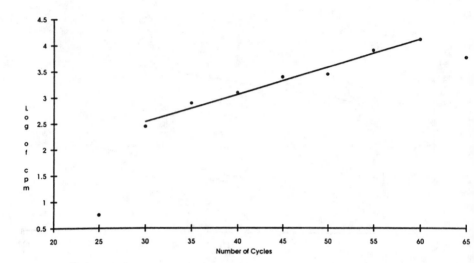

Fig. 2. Ten nanograms of RNA extracted from paraffin-embedded tissues are reverse transcribed and amplified from 25 to 65 cycles for β-actin. The linearity is preserved between 30 and 60 cycles.

Table 1
Standardization of RNA Degradation Level
in EGFr mRNA Quantitation in Brain Gliomas

Samples	cpm β-actin[a]	Standardization factor[b]	cpm EGFr[a]	Standardized EGFr[c]
1	143	6.804196	61	415.0
2	299	3.254181	265	862.3
3	1404	0.69302	662	458.7
4	1582	0.615044	468	287.8
5	1082	0.899261	883	794.0
6	816	1.192402	308	367.2
7	1151	0.845352	284	240.0
8	867	1.122261	288	323.2
9	1063	0.915334	135	123.5
10	1322	0.736006	86	63.2

[a] Counts per minute of the membrane cut from dot-blot amplified for β-actin and EGFr after empty membrane cpm subtraction. In this case, β-actin mRNA is used as reference sequence.

[b] Standardization factor for each sample calculated as ratio between the average, β-Actin value and cpm of β-actin for each sample.

[c] EGFr results after standardization (product of values in column 3 times values in column 4).

different levels of RNA degradation among samples. For a precise comparison, the results must be standardized on the basis of quantization of a reference mRNA well expressed in every tissue like that of β-actin or GAPDH. To standardize the data, we calculate the mean value of the cpm of the reference mRNA for all the cases studied. The mean value is then divided by the results of the cpm for the reference sequence of each sample and the ratio multiplied by the counts of the specific mRNA to compare (**Table 1**, previous page). For quantification purposes we use common master-mixes for all the steps of the case study analysis.

4. Notes

1. The spectrophotometer quantitation at 260 nm ([RNA] $\mu g/\mu L$ = Absorbance$_{260}$ × 40×10^{-3} × Dilution factor) is not accurate because of the high dilution of the RNA samples obtained from paraffin-embedded tissues. But an approximate quantification of the total RNA extracted does not affect the method because of the standardization of the results.

2. To be sure that all the product is applied on the membrane we do not use the upper part of the apparatus; in fact, when the vacuum is open the membrane adheres directly to the lower part of the apparatus. To facilitate the cut of the membrane for scintillation counting, it is better to spot the samples every second hole, leaving an empty line above and below each sample.

References

1. Stanta, G. and Schneider, C. (1991) RNA extracted from paraffin-embedded human tissues is amenable to analysis by PCR amplification. *BioTechniques* **11**, 304–308.

2. Maniatis, T., Fritsch, E. F., and Sambrook, J. (1989) *Molecular Cloning: A Laboratory Manual.* (2nd ed.). Cold Spring Habor Laboratory, Cold Spring Harbor, NY.

3. Stanta, G. and Bonin, S. RNA Quantitative analysis from fixed and paraffin-embedded tissues: membrane hybridization and capillary electrophoresis comparison. *BioTechniques,* in press.

19

Quantitative Analysis of RNA Species by PCR and Solid-Phase Minisequencing

Anu Suomalainen and Ann-Christine Syvänen

I. Introduction

Methods allowing sensitive and accurate quantitative analysis of defined RNA species are required in a wide variety of gene expression studies. Unlike the traditional hybridization methods, RNase protection or S1 nuclease assays (see Chapters 16 and 27), the methods based on reverse transcription (RT) and PCR provide an essentially unlimited sensitivity of detection. A drawback of the PCR-based methods is, however, that they do not allow direct quantification of a sequence present in a sample, because the efficiency of the PCR depends on the amount of the template sequence, and the amplification is exponential only at low concentrations of the template (1). Due to this "plateau effect" of the PCR, the amount of the amplification product does not reflect directly the original amount of the template. Moreover, subtle differences in the reaction conditions may cause a significant amount of sample-to-sample variation in the final yield of the PCR. The efficiency of amplification is also affected by the sequence of the PCR primers, as well as the size and, to some extent, the sequence of the PCR product. For these reasons a prerequisite for an accurate quantitative PCR analysis is that an internal standard is coamplified in the same reaction with the target sequence. The standard sequence should be as similar to the target sequence as possible to ensure that the target-to-standard ratio remains constant throughout the amplification. An ideal PCR standard differs from the target sequence only at one nucleotide position, by which the two sequences can be identified and quantified after the amplification. For quantification of RNA the optimal internal standard is RNA, not only to control the efficiency of the PCR, but also that of the cDNA synthesis.

From: Methods in Molecular Biology, Vol. 86: RNA Isolation and Characterization Protocols
Edited by: R. Rapley and D. L. Manning © Humana Press Inc., Totowa, NJ

The determination of the relative amounts of the PCR products originating from the target and standard sequences allows calculation of the initial amount of the target sequence *(2–4)*. For many applications it is sufficient to determine the relative amount of the target sequence compared to an endogenous standard sequence originally present in the sample. To be able to determine the absolute amount of a target sequence present, it is necessary to add a known amount of a standard sequence to the sample before amplification. In this case a measure of the amount of the analyzed sample, such as the number of cells or total amount of RNA, is needed. Finally, an accurate method for detecting the PCR products originating from the standard and target sequences is required.

We have developed the solid-phase minisequencing method for the detection and quantification of nucleotide sequence variants differing from each other at one nucleotide position *(5, 6)*. In this method a DNA fragment spanning the site of the variable nucleotide is first amplified using one biotinylated and one unbiotinylated PCR primer. The PCR product carrying biotin at the 5' end of one of its strands is captured on an avidin-coated solid support and denatured. The nucleotides at the variable site in the immobilized DNA strands are identified by two separate primer extension reactions, in which a single labeled deoxynucleotide triphosphate (dNTP) is incorporated by a DNA polymerase. In our standard format of the assay, [³H]dNTPs serve as labels and streptavidin-coated microtiter plates are used as the solid support *(6)*. The results of the assay are numeric counts per minute (cpm) values expressing the amount of the specific [³H]dNTPs incorporated in the minisequencing reactions. When analyzing samples containing a mixture of two sequences differing from each other by a single nucleotide, the ratio between the two cpm values reflects directly the ratio between the two sequences in the original sample. Because the two sequences are essentially identical, they are amplified with equal efficiency irrespective of the exponential phase of the PCR. The solid-phase format of the assay allows interpretation of the result by direct measurement of the incorporated label without any separation step that could introduce errors into the analysis. These features make the solid-phase minisequencing method close to an ideal tool for PCR-based determination of both relative and absolute amounts of DNA and RNA sequences. **Figure 1** illustrates the principle of the quantitative analysis by solid-phase minisequencing. This method has been applied, for example, to accurately compare the transcript levels of mutant and normal alleles of one gene and of two highly homologous genes *(7)*, as well as to accurately determine the absolute amounts of specific transcripts in cell and tissue samples utilizing an internal synthetic RNA *(8)* or DNA standard *(9)*, designed to differ from the target RNA of interest only by one nucleotide. **Table 1** summarizes the alternative strategies and how the solid-phase minisequencing method has been utilized for quantitative PCR analysis of RNA.

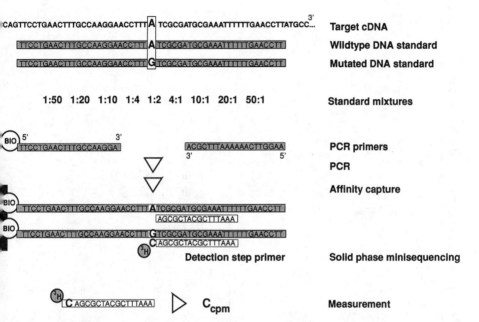

Fig. 1. Quantitative analysis by solid-phase minisequencing utilizing oligonucleotide standards. 1. Design of DNA standards (oligonucleotides). The wild-type DNA standard is identical to the target cDNA sequence, and the mutated standard differs from it by one nucleotide. The standards are mixed in known ratios. 2. PCR. The 5' PCR primer carries a biotin residue (BIO). This results in a PCR product carrying biotin in the 5' end of one of its strands. 3. Solid-phase minisequencing analysis. The analysis for one of the nucleotides (G, the mutated nucleotide) is shown. The PCR product is captured in a streptavidin-coated microtiter well and denatured. A detection step primer hybridizes to the single-stranded template, 3' adjacent to the variant nucleotide. The DNA polymerase extends the primer with the [³H]-labeled dNTP, if it is complementary to the nucleotide present at the variable site. 4. Denaturation and measurement of the fluted radioactivity.

2. Materials

2.1. Equipment

1. Programmable heat block and facilities to avoid contamination in PCR.
2. Microtiter plates with streptavidin-coated wells (e.g., Combiplate 8, Labsystems, Helsinki, Finland) (*see* **Note 1**).
3. Multichannel pipet and microtiter plate washer (optional).
4. Shaker at 37°C.
5. Water bath or incubator at 42°C and 50°C.
6. Liquid scintillation counter.

Table 1
Summary of Alternative Strategies for Quantitative PCR Analysis of RNA by Solid-Phase Minisequencing

Relative RNA quantification	Application	Interpretation	Ref.
Standard			
Two variants of a transcript present as a mixture in a sample	Screening for differences in levels of transcripts of two alleles or two homologous genes.	1. Calculate the result from the ratio between the incorporated [^{3}H]dNTPs after correction for differences in specific activities and number of incorporated dNTPs	6,9
Serve as standards for each other during RT and PCR			
Genomic DNA		2. Calculate the result by comparing the ratio between the incorporated [^{3}H]dNTPs with that obtained from the genomic DNA	7
Serves as an external reference for the sequence variant in a natural 1:2 ratio		3. Calculate the result from a complete standard curve constructed with mixtures of the two sequence variants	

Absolute RNA quantification	Application	Interpretation	Ref.
Standard			
RNA fragment	Determination of the absolute amount of target RNA in a defined sample. Both the RT and PCR are controlled.	As 1 or 3 above.	8
Differs from the target at one nucleotide			
Prepared by in vitro transcription from a DNA construct (in the future by chemical synthesis)			
Mixed with the target RNA before RT			
DNA fragment	Determination of the absolute amount of the target sequence in a defined sample. Does not control the RT reaction.	As 1 or 3 above.	6
Differs from the target at one nucleotide			
Oligonucleotide, cloned DNA or PCR-product			
Mixed with the template			

2.2. Reagents

2.2.1. Reverse Transcription (store all reagents at –20°C)

1. Avian myeloblastosis virus (AMV) reverse transcriptase (20–25 U/µL, Promega, Madison, WI).
2. RNasin ribonuclease inhibitor (20–40 U/µL, Promega).
3. 10X concentrated RT buffer: 500 mM Tris-HCl, pH 8.2–8.5, 50 mM MgCl$_2$, 400 mM KCl, and 20 mM dithiothreitol.
4. dNTP mixture: 2 mM dATP, 2 mM dCTP, 2 mM dGTP, and 2 mM dTTP.
5. Diethyl pyrocarbonate treated water (DEPC H$_2$O).

2.2.2. PCR (store at –20°C)

1. Thermostable DNA polymerase. We use *Thermus aquaticus* (5 U/µL, Promega) or *Thermus brockianus* (Dynazyme II, 2 U/µL, Finnzymes, Espoo, Finland) DNA polymerase (*see* **Note 2**).
2. 10X concentrated DNA polymerase buffer: 500 mM Tris-HCl, pH 8.8, 150 mM (NH$_4$)$_2$SO$_4$, 15 mM MgCl$_2$, 1% v/v Triton X-100, 0.1% w/v gelatin.
3. dNTP mixture as in **Step 4., Subheading 2.2.**

2.2.3. Minisequencing Analysis

1. PBS/Tween: 20 mM sodium phosphate buffer, pH 7.5, and 0.1% (v/v) Tween-20. Store at 4°C. 50 mL is enough for several full-plate analyses.
2. TENT (washing solution): 40 mM Tris-HCl, pH 8.8, 1 mM EDTA, 50 mM NaCl, and 0.1% (v/v) Tween-20. Store at 4°C. Prepare 1–2 L at a time, which is enough for several full-plate analyses.
3. Use 50 mM NaOH (make fresh every 4 wk), store at room temperature (approx 20°C). Prepare 50 mL.
4. [^{3}H]-labeled deoxynucleotides (dNTPs): dATP to detect a T at the variant site, dCTP to detect a G, and so on. (Amersham; [^{3}H]dATP, TRK 625; dCTP, TRK 576; dGTP, TRK 627; dTTP, TRK 633), store at –20°C (*see* **Note 3**).
5. Scintillation reagent (for example, Hi-Safe II, Wallac, Turku, Finalnd).

2.3. Primer Design

1. PCR primers: Biotinylate one of the PCR primers at its 5' end during the synthesis using a biotin-phosphoramidite reagent (for example, Amersham, UK or Perkin Elmer/ABI, Forster City, CA) (*see* **Note 4**). The 3' primer can be used for the RT, and as the 3' PCR primer. If oligonucleotides (DNA) are used as quantification standards, the length of oligonucleotides that can be synthesized with acceptable yields sets an upper limit of about 80–100 base pairs for the PCR product.
2. Detection step primer for the minisequencing analysis is an oligonucleotide complementary to the biotinylated strand, designed to hybridize with its 3' end with the nucleotide adjacent to the variant nucleotide to be analyzed (*see* **Fig. 1**). The detection step primer for our standard protocol is a 20-mer. The primer should be at least five nucleotides nested in relation to the unbiotinylated PCR primer.

2.4. Quantification Standards

The standard should be designed to differ from the target sequence by one nucleotide, which is the one detected in the minisequencing analysis (*see* **Fig. 1**). If a standard curve is to be constructed, a second standard, identical to the target sequence is required (see **Subheading 3.4.**). Depending on the application, either DNA or RNA standards may be used (*see* **Table 1**). RNA standards are synthesized by in vitro transcription *(10)* and oligonucleotide standards using a DNA/RNA synthesizer (*see* **Note 5**). PCR products or cloned cDNA fragments can also be used as standards. Measure the molecular concentrations of the standards. The optimal amount of the standard added to a sample depends on the abundance of the target sequence in the original sample. The ratio of the target to the standard sequence should preferably be between 0.1 to 10. If no estimate of the amount of the target sequence is available, it may be necessary to initially titrate the optimal amount of the standard using several amounts of the standard in the analysis (for example, 10^2, 10^4, 10^6, and 10^8 molecules). For accurate quantification, standards representing both sequence variants should be available, and analysis of mixtures of known amounts of the two standards should be analyzed to construct a standard curve, as demonstrated in **Figs. 1** and **2**.

3. Methods

3.1. Reverse Transcription

Any established protocol for RT can be applied. We use the following protocol:

1. Prepare an RT mixture by combining 2 µL of 10X RT buffer, 5 µL of dNTP mixture, 50 pmol of the 3' PCR primer, 10 U RNasin, 30 U AMV reverse transcriptase, and DEPC H_2O to a final volume of 15 µL. Keep on ice until used.
2. Add the appropriate amount of RNA (usually 50–500 ng) in 5 µL depH_2O to an Eppendorf tube on ice. If an RNA standard is used, add the standard to the sample at this stage, keeping the total volume of the RT reaction at 20 µL. Proceed immediately to the next step.
3. Add 15 µL of the RT mixture to the RNA sample.
4. Incubate for 60 min at 42°C. The samples can be stored at −20°C for 1–3 d.

3.2. PCR for Solid-Phase Minisequencing Analysis

The PCR follows the routine protocols, except that the amount of the biotin-labeled primer should be reduced not to exceed the biotin-binding capacity of the microtiter well (*see* **Note 1**). For a 50 µL PCR reaction we use 10 pmol of biotin-labeled primer and 50 pmol of the unbiotinylated primer, and as template, $^1/_4$ of the RT product. The PCR should be optimized (i.e., the annealing temperature and the amount of the template) to be efficient and specific. To be able to use [^{3}H]dNTPs, which have low specific activities, for

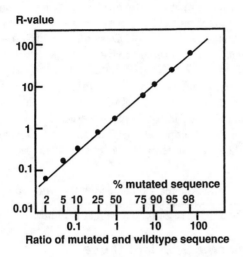

Fig. 2. Standard curve obtained by plotting the R values (C_{cpm}/T_{cpm}) as a function of the ratio between the standard sequences on a log–log scale.

the minisequencing analysis, $1/10$ of the PCR product should produce a single visible band after agarose gel electrophoresis and staining with ethidium bromide (*see* Chapter 13).

3.3. Solid-Phase Minisequencing Analysis

1. Affinity capture: Transfer 10 µL aliquots of the PCR product and 40 µL of PBS/ Tween to two streptavidin-coated microtiter wells (*see* **Note 6**). Include as negative controls two wells without PCR product. Seal the wells with a sticker and incubate the plate at 37°C for 1.5 h with gentle shaking.
2. Discard the liquid from the wells and tap the wells dry against tissue paper.
3. Wash the wells three times at room temperature by adding 200 µL of TENT to each well, discard the washing solution and empty the wells thoroughly between the washing steps (*see* **Note 7**).
4. Denature the captured PCR product by adding 100 µL of 50 m*M* NaOH to each well, followed by incubation at room temperature for 3 min. Discard the NaOH and wash the wells as in **Subheading 3.3., step 3**.
5. For each DNA fragment to be analyzed prepare two 50 µL mixtures of nucleotide-specific minisequencing solution, one for detection of the wild-type and one for the mutant nucleotide, by mixing 5 µL of 10X DNA polymerase buffer, 10 pmol of detection step primer (for example, 2 µL of 5 µ*M* primer), 0.2 µCi (usually 0.2 µL) of one [³H]dNTP, 0.1 U of DNA polymerase, and H₂O to a total volume of 50 µL. It is obviously convenient to prepare master mixes for the desired number of analyses with each nucleotide (*see* **Note 8**).

6. Add 50 µL of one nucleotide-specific mixture to each well, and incubate the plate at 50°C for 10 min (*see* **Note 9**).
7. Discard the contents of the wells and wash them as in **Subheading 3.3., step 3**.
8. Release the detection step primer from the template by adding 60 µL 50 m*M* NaOH and incubating for 3 min at room temperature.
9. Transfer the NaOH containing the eluted primer to the scintillation vials, add scintillation reagent, and measure the radioactivity, i.e., the amount of incorporated label, in a liquid scintillation counter *(see* **Note 10**).
10. The result is obtained as cpm values. The cpm value of each reaction expresses the amount of the incorporated [³H]dNTP. Calculate the ratio (R) (*see* **Table 2** and **Note 11**):

$$R = \frac{\text{cpm incorporated in the reaction detecting the wild-type (target) sequence}}{\text{cpm incorporated in the reaction detecting the mutated (standard) sequence}} \quad (1)$$

3.4. Preparation of the Standard Curve

Mix the wild-type and mutated standard sequences in known proportions, for example, 1:50, 1:20, 1:10, 1:4, 1:2, 4:1, 10:1, 20:1, and 50:1. If RNA standard is used, synthesize cDNA using the mixtures as the template, amplify the product by PCR and analyze the products by minisequencing. Plot the resulting R values on a log–log scale as a function of the ratio between the sequences present in the original mixture, which should result in a linear standard curve (*see* **Table 2** and **Fig. 2**). This curve can then be utilized for the analysis of the actual samples with an unknown amount of target RNA either to determine the relative or the absolute amount of the target RNA (*see* **Table 1**).

4. Notes

1. The binding capacity of the streptavidin-coated microtiter well that we use is 2–5 pmol of biotinylated oligonucleotide. If a higher binding capacity is desired, avidin-coated polystyrene beads (Fluoricon, 0.99 µm, IDEXX, Portland ME; biotin-binding capacity over 2 nmol of oligonucleotide/mg beads) or streptavidin-coated magnetic polystyrene beads (Dynabeads M-280, streptavidin [Dynal, Oslo, Norway]; biotin-binding capacity 300 pmol/mg) can be used *(11)*. The biotinbinding capacity of a microtiter well allows reliable detection of approx 2% of a sequence-variant present in a sample *(9)*, whereas a detection sensitivity of < 0.1% can be obtained with the bead-based format *(12)*.
2. The use of a thermostable DNA polymerase in the single-nucleotide primer extension reaction is advantageous since a high temperature, also favorable for the simultaneous primer annealing reaction, can be used.
3. Although the [³H]dNTPs are weak β-emitters, their half lives are long (13 yr), and the necessary precautions for working with [³H] should be taken. Also dNTPs or dideoxynucleotides labeled with other isotopes ([³⁵S] or [³²P], **ref. 5**) or with fluorophores *(13)* can be used.

**Table 2
Example of the Result of a Solid-Phase Minisequencing
Analysis of Mixtures of Two Standard Oligonucleotides**

Ratio of wildtype sequence to mutated sequence	T_{cpm} (wt oligo)[a]	C_{cpm} (mut oligo)[b]	R value C_{cpm}/T_{cpm}
Wildtype	3110	44	0.014
50:1	3640	190	0.05
20:1	2780	420	0.15
10:1	2830	730	0.26
4:1	2520	1690	0.67
1:2	1650	2810	1.7
1:4	790	3630	4.6
1:10	350	3790	10.8
1:20	210	4760	22.7
1:50	120	4800	40.0
Mutant	43	4580	106.5
H_2O	41	23	—

[a] The specific activities of the [³H]dNTPs: dTTP 126 Ci/mmol, dCTP 67 Ci/mmol.
[b] In this case, two [³H]dCTPs were incorporated into the mutant sequence.

4. The efficiency of the 5'-biotinylation of an oligonucleotide on a DNA synthesizer is most often 80%–90%. The biotin-labeled oligonucleotides can be purified from the unbiotinylated ones either by high-performance liquid chromatography *(14)*, polyacrylamide gel electrophoresis *(15)* or by disposable ion exchange columns manufactured for this purpose (Perkin-Elmer/ABI). If the biotin-labeled primer is used without purification, the success of the biotinylation can be confirmed after the PCR by affinity capture of an aliquot of the biotinylated PCR product on an avidin matrix with high biotin-binding capacity (*see* **Note 1**). Analyze the supernatant after the capturing reaction by agarose gel electrophoresis. If the biotinylation has been efficient, no product, or a faint product of significantly lower intensity than the unbound PCR product is observed in the supernatant.

5. For use as quantification standards, full-length oligonucleotides should be purified from prematurely terminated ones by high-performance liquid chromatography *(14)* or by sizeseparation in polyacrylamide gel electrophoresis *(15)*. The molecular concentration of the purified full-length standard DNA or RNA can then be accurately determined. At present it is not feasible to produce RNA fragments chemically by a DNA/RNA synthesizer to serve as RNA standards in a quantitative analysis.

6. Each nucleotide to be detected at the variant site is analyzed in a separate well. Thus, at least two wells are needed per PCR product. For quantitative applications we carry out two parallel assays for each nucleotide, i.e., four wells per PCR product.

7. The washing can be performed utilizing an automated microtiter plate washer, or by manually pipeting the washing solution to the wells, discarding the liquid and tapping the plate against a tissue paper. It is important to thoroughly empty the wells between the washing steps.

8. The minisequencing reaction mixture can be stored at room temperature for 1–2 h. It is convenient to prepare it during the incubation in **Subheading 3., step 1**.

9. The conditions for hybridizing the detection step primer are not stringent, and the temperature of 50°C can be used to analyze of most PCR products irrespectively of the sequence of the detection step primer. If the primer, however, is considerably shorter than 20-mer or if its GC content is low (melting temperature close to 50°C) lower temperatures for the primer annealing may be required.

10. Streptavidin-coated microtiter plates made of scintillating polystyrene are available (ScintiStrips, Wallac, Finland). When these plates are used, the final washing, denaturation, and transfer of the fluted detection primer to scintillation vials can be omitted, but a scintillation counter for microtiter plates is needed *(16)*.

11. The ratio between the cpm values for the two nucleotides reflects the ratio between the two sequences in the original sample. The R value is affected by the specific activities of the [³H]dNTPs used, and if either the wildtype or the mutant sequence allows the detection step primer to be extended by more than one [³H]dNTP, this will obviously also affect the R value. Both of these factors can easily be corrected, when calculating the ratio between the two sequences. Another possibility is to construct a standard curve (*see* **Subheading 3.4.**).

References

1. Syvänen, A.-C., Bengtström, M., Tenliunen, J., and Söderlund, H. (1988) Quantification of polymerase chain reaction products by affinity-based hybrid collection. *Nucl. Acids Res.* **16**, 11,327–11,338.

2. Chelly, J., Kaplan, J.-C., Maire, P., Gautron, S., and Kahn, A. (1988) Transcription of the dystrophin gene in human muscle and non-muscle tissues. *Nature* **333**, 858–860.

3. Wang, A. M., Doyle, M. V., and Mark, D. F. (1989) Quantitation of mRNA by the polymerase chain reaction. *Proc. Natl. Acad. Sci. USA* **86**, 9717–9721.

4. Gilliland, G., Perrin, S., Blanchard, K., and Bunn, H. F. (1990) Analysis of cytokine mRNA and DNA: detection and quantitation by competitive polymerase chain reaction. *Proc. Natl. Acad. Sci. USA* **87**, 2725–2729.

5. Syvänen, A.-C., Aalto-Setälä, K., Harju, L., Kontula, K., and Söderlund, H. (1990) A primer-guided nucleotide incorporation assay in the genotyping of apolipoprotein E. *Genomics* **8**, 684–692.

6. Syvänen, A.-C., Sajantila, A., and Lukka, M. (1993) Identification of individuals by analysis of biallelic DNA markers, using PCR and solid-phase minisequencing. *Am. J. Hum. Genet.* **52**, 46–59.

7. Karttunen, L., Lönnqvist, L., Godfrey, M., Peltonen, L., and Syvänen, A.-C. (1996) An accurate method for comparing transcript levels of two alleles or highly homologous genes: application to fibrillin transcripts in Marfan patients' fibroblasts. *Genome Res.* **6**, 392–403.

8. Ikonen, E., Manninen, T., Peltonen, L., and Syvänen, A.-C. (1992) Quantitative determination of rare mRNA species by PCR and solid-phase minisequencing. *PCR Methods Appl.* **1**, 234–240.

9. Suomalainen, A., Majander, A., Enihko, H., Peltonen, L., and Syvänen, A.-C. (1993) Quantification of tRNA$_{3243}^{Leu}$ point mutation of mitochondrial DNA in MELAS patients and its effects on mitochondrial transcription. *Hum. Molec. Genet.* **2**, 525–534.

10. Melton, D. A., Krieg, P. A., Rebagliati, M. R., Maniatis, T., Zinn, K., and Green, M. R. (1984) Efficient in vitro synthesis of biologically active RNA and RNA hybridization probes from plasmids containing a bacteriophage SP6 promoter. *Nucleic Acids Res.* **12**, 7035–7056.

11. Syvänen, A.-C. and Söderlund, H. (1993) Quantification of polymerase chain reaction products by affinity-based collection. *Meth. Enzymol.* **218**, 474–490.

12. Syvänen, A.-C., Söderlund, H., Laaksonen, E., Bengtström, M., Turunen, M., and Palotie, A. (1992) N-ras gene mutations in acutle myeloid leukemia: accurate detection by solid-phase minisequencing. *Int. J. Cancer* **50**, 713–718.

13. Pastinen, T., Partanen, J., and Syvänen, A.-C. (1996) Multiplex, fluorescent solid-phase minisequencing for efficient screening of DNA sequence variation. *Clin. Chem.* **42**, 1391–1397.

14. Bengtström, M., Jungell-Nortamo, A., and Syvänen, A.-C. (1990) Biotinylation of oligonucleotides using a water soluble biotin ester. *Nucleosides Nucleotides* **9**, 123–127.

15. Wu, R., Wu, N. -H., Hanna, Z., Georges, F., and Narang, S. (1984) In: *Oligonucleotide Synthesis: A Practical Approach.* (Ed. Gait, M. J.), IRL Press, Oxford, UK, p. 135.

16. Ihalainen, J., Siitari, H., Laine, S., Syvänen, A. -C., and Palotie, A. (1994) Towards automatic detection of point mutations: use of scintillating microplates in solid-phase minisequencing. *BioTechniques* **16**, 938–943.

20

Preparation of Tissue Sections and Slides for mRNA Hybridization

Giorgio Terenghi

1. Introduction

The first step for a successful *in situ* hybridization is the fixation of the tissue. This will ensure target nucleic acid retention and preservation of the tissue morphology. Either crosslinking or precipitative fixatives can be used, and a preference for either of the two types of fixative has often been based on the different types of system under investigation *(1–7)*. For hybridization of regulatory peptide mRNA, 4% paraformaldehyde appears to be the most effective, both on tissue blocks and on tissue culture preparations.

When manipulating tissue to be used for mRNA hybridization, it is essential to work in RNase-free conditions. RNase is an ubiquitous and heat-resistant enzyme that degrades any single-stranded RNA molecule very rapidly. Fingertips are particularly rich in ribonuclease, hence clean, disposable gloves should be worn at all times. All equipment and solutions should also be RNase-free. Fixative solutions, as they exert an inhibitory action on this enzyme, are naturally RNase-free.

Fixed tissue is generally processed for cryostat sectioning, but *in situ* hybridization can be equally successful on paraffin embedded tissue *(8,9)*. In any case, it is important always to keep the delay between tissue collection and fixation to a minimum, to avoid nucleic acid degradation, which obviously increases with time delay *(4,10)*. This is not a problem when using experimental animal tissue or cell culture, but it becomes an extremely important point when using surgical or postmortem material. As the degradation curve varies for different mRNAs, there is no fixed rule on an acceptable time limit, but a delay of 15–30 min is considered acceptable in most cases.

From: *Methods in Molecular Biology, Vol. 86: RNA Isolation and Characterization Protocols*
Edited by: R. Rapley and D. L. Manning © Humana Press Inc., Totowa, NJ

The fixed material can be stored in washing buffer only for a limited time (up to 1 mo), but frozen or paraffin tissue blocks can be safely stored for many months or years. Cryostat blocks should be stored at −40°C or below, and room temperature is considered adequate for paraffin blocks. Cryostat sections can be stored dry at −70°C for up to 1 yr, and wax sections keep at room temperature (dewaxed sections can be kept in 70% alcohol at 4°C), given that RNase-free conditions are observed.

Tissue sections should be collected on Vectrabond (Vectra, UK) coated slides to prevent loss of material during the many steps of the hybridization procedure. The best tissue adhesion is obtained if the sections are left to dry for at least 4 h (or overnight) at 37°C before use or storage.

2. Materials

Plastic disposable equipment and solutions should be autoclaved before use. RNase inhibitors (i.e., human placental ribonuclease inhibitor, DEPC, heparin, and so on) can be added to solutions containing enzymes, which are heat sensitive.

1. Phosphate buffered saline (PBS): Dissolve in 9 L of distilled water the following: 87.9 g NaCl, 2.72 g KH_2PO_4, 11.35 g anhydrous Na_2HPO_4. Adjust the pH to 7.1–7.2 with HCl before adjusting the total volume to 10 L. This solution can be stored at room temperature ready for use.
2. 4% Paraformaldehyde solution: Dissolve 4 g of paraformaldehyde in 80 mL of 0.01M PBS with heat, keeping the temperature below 60°C. Stir the slurry until the powder is completely dissolved, and if necessary add few drops of 10N NaOH to clear the solution. Adjust the volume to 100 mL and leave the solution to cool before using. The fixative should be freshly prepared before use.
3. Vectrabond: Prepare the solution according to the manufacturer instructions, and use on RNase-free coated slides.
4. RNase-free glass slides: Place some slides in metal racks and soak them in 70% ethanol for 10 min before proceeding with the coating according to manufacturer instructions. Always wear gloves during this procedure.
5. Liquid nitrogen.
6. Isopentane.
7. Freezing glue (e.g., Tissue-Tek OCT, Miles, New Haven, CT).
8. Washing buffer: 15% (w/v) sucrose, 0.01% (w/v) sodium azide dissolved in PBS.
9. Plastic disposable equipment and solutions should be autoclaved before use. RNase inhibitors (i.e., human placental ribonuclease inhibitor, DEPC, heparin, and so on) can be added to solutions containing enzymes that are heat sensitive.

3. Methods
3.1. Fixation (see Note 2)

1. Cut tissue into small pieces (approx 1 × 1 × 0.5 cm) using a sterile sharp blade (*see* **Note 2**).

2. Fix the tissue in freshly made 4% paraformaldehyde solution for 6 h at room temperature (*see* **Notes 3** and **4**).
3. After fixation, rinse the tissue blocks in four to five changes of washing buffer (2 h each change or overnight).
4. Store the fixed material in washing buffer at 4°C in labeled containers, ready for cryostat or wax blocking (*see* **Note 5**).

3.2. Preparing the Cryostat Block and Sectioning Tissue

1. Precool a Pyrex or metal beaker by immersion in liquid nitrogen and fill it with Arcton or isopentane.
2. Immerse the beaker again in liquid nitrogen and freeze the Arcton (isopentane), then remove the beaker from the flask and allow the Arcton (isopentane) to melt until there is enough liquid to cover the block, but still some solid in the bottom to maintain the temperature as low as possible.
3. Mount the tissue on a cork disk and surround it with special freezing glue (e.g., OCT) (*see* **Notes 6** and **7**).
4. Hold the cork disk with forceps and lower it quickly in the melting Arcton (isopentane) to snap-freeze the tissue.
5. Transfer the frozen block in precooled plastic bag for storage in liquid nitrogen or at –40°C.
6. The block should be allowed to warm up to cryostat temperature (–20°C) before cutting is attempted.
7. Mount the block on the cryostat head and cut thin sections (10–30 mm), picking them up onto PLL-coated slides (*see* **Notes 1, 5,** and **7**).
8. Dry the sections for at least 4 h (or overnight) at 37°C to obtain maximum tissue adhesion.

4. Notes

1. Batches of coated slides may be prepared in advance and stored in racks, wrapped in the aluminum foil to protect from dust and RNase contamination. Vetrabond-coated slides can be stored at room temperature for up to 1 mo.
2. When mRNA is the target, special care should be taken, and all the specimen handling procedures should be carried out using clean, disposable gloves and sterile instruments, in order to avoid RNase contamination.
3. Animal tissue can be fixed *in situ* by perfusion with 4% paraformaldehyde, followed by immersion fixation of the dissected tissue (1–4 h, depending on the tissue size and fixation obtained with perfusion). This method is strongly recommended if brain or spinal cord tissues are used, as these tissues do not fix well by immersion only, owing to the poor penetration of the fixative in the tissue matrix.
4. Fixative other than paraformaldehyde may be more appropriate when investigating specific target molecules. Some indication of other possible fixatives can be gained from the listed literature. However, it is good practice to test more than one fixative in order to establish which give best hybridization results, as

indicated by the highest signal:background noise ratio with optimal preservation of morphology.

5. During storage of the cryostat blocks, do not leave any tissue surface uncovered, as it will dry out and it will become impossible to cut. After cutting, spread a thin layer of OCT glue on the cut surface of the block, and leave at –20°C until frozen. Store the block in a sealed plastic bag or other appropriate container.

6. Tissue blocks should be orientated so that the face you wish to cut is uppermost. Cryostat blocks cut best in the vertical plane, unlike paraffin blocks.

7. Very small pieces of tissue should be mounted on another piece of inert or inappropriate tissue (e.g., liver), suitably trimmed, so that adequate clearance is obtained on cutting.

References

1. Haase, A. T., Brahic, M., and Stowring, L. (1984) Detection of viral nucleic acids by *in situ* hybridization, in *Methods in Virology,* vol. VII (Maramorosch, K. and Koprowski, H., eds.), Academic, New York, pp. 189–226.

2. McAllister, H. A. and Rock, D. L. (1985) Comparative usefulness of tissue fixatives for *in situ* viral nucleic acid hybridization. *J. Histochem. Cytochem.* **33,** 1026–1032.

3. Moench, T. R., Gendelman, H. E., Clements, J. E., Narayan, O., and Griffin, D. E. (1985) Efficiency of *in situ* hybridization as a function of probe size and fixation technique. *J. Virol. Method.* **11,** 119–130.

4. Hofler, H., Childers, H., Montminy, M. R., Lechan, R. M., Goodman, R. H., and Wolfe, H. J. (1986) *In situ* hybridization methods for the detection of somatostatin mRNA in tissue sections using antisense RNA probes. *Histochem. J.* **18,** 597–604.

5. Singer, R. H., Lawrence, J. B., and Villnave, C. (1986) Optimization of *in situ* hybridization using isotopic and non-isotopic detection methods. *BioTechniques* **4,** 230–250.

6. Guitteny, A. F., Fouque, B., Mongin, C., Teoule, R., and Boch, B. (1988) Histological detection of mRNA with biotinylated synthetic oligonucleotide probes. *J. Histochem. Cytochem.* **36,** 563–571.

7. Terenghi, G. and Fallon, R. A. (1990) Techniques and applications of *in situ* hybridization, in *Current Topics in Pathology: Pathology of the Nucleus* (Underwood, J. C. E., ed.), Springer Verlag, Berlin, pp. 290–337.

8. Farquharson, M., Harvie, R., and McNicol, A. M. (1990) Detection of mRNA using a digoxigenin end labelled oligodeoxynucleotide probe. *J. Clin. Pathol.* **43,** 424–428.

9. Unger, E. R., Hammer, M. I., and Chenggis, M. L. (1991) Comparison of 35S and biotin as labels for *in situ* hybridization: use of an HPV model system. *J. Histochem. Cytochem.* **39,** 145–150.

10. Asanuma, M., Ogawa, N., Mizukawa, K., Haba, K., and Mori, A. (1990) A comparison of formaldehyde-preperfused frozen and freshly frozen tissue preparation for the *in situ* hybridization for α-tubulin mRNA in the rat brain. *Res. Comm. Chem. Pathol. Pharmacol.* **70,** 183–192.

21

Detecting mRNA in Tissue Sections with Digoxigenin-Labeled Probes

Giorgio Terenghi

1. Introduction

Nonradioactively labeled probes offer several advantages compared to radioactive ones, as they show long stability, high morphological resolution, and rapid developing time. There are different types of nonradioactive labeling methods available, although digoxigenin-labeled probes *(1)* have become the most widely used for investigation on animal tissue, as they offer the advantage of low background noise and increased sensitivity *(2,3)*. Also, digoxigenin can be used to label either RNA, DNA, or oligonucleotide probes. There have been different opinions on the sensitivity of detection of digoxigenin probes, but it has been shown that the sensitivity of radiolabeled and nonradioactive probes is comparable *(3)*.

The detection of digoxigenin-labeled probes is carried out with immunohistochemical methods, using antidigoxigenin antibodies that are conjugated to either fluorescent or enzymatic reporter molecules *(1)*. However, it has to be remembered that different immunohistochemical detection systems might determine the resolution and detection sensitivity of *in situ* hybridization *(4)*. The variety of detection methods also offers the possibility to carry out double *in situ* hybridization, e.g., using digoxigenin-labeled probes and directly labeled probes, which are then visualized using different immunohistochemical methods *(5)*. Alternatively, digoxigenin and biotin can be used in combination for the identification of two different nucleic acid sequences on the same sections *(6,7)*.

From: *Methods in Molecular Biology, Vol. 86: RNA Isolation and Characterization Protocols*
Edited by: R. Rapley and D. L. Manning © Humana Press Inc., Totowa, NJ

2. Materials

1. 1M Tris-HCl, pH 8.0: Dissolve 121.1 g Tris-base in 800 mL double-distilled water. Adjust to pH 8.0 with HCl, then adjust volume to 1 L with double-distilled water before autoclaving.

2. 0.5M EDTA: Add 186.1 g of $Na_2EDTA \cdot 2H_2O$ to approx 600 mL double-distilled water. Stir continuously keeping the solution at 60°C, adding NaOH pellets (approx 20 g) until near pH 8.0. Only then will the EDTA start to dissolve. When completely dissolved, leave the solution to cool down to room temperature, then adjust to pH 8.0 with 10N NaOH solution. Adjust the volume to 1 L with double-distilled water and autoclave.

3. 1M Glycine: Dissolve 75 g glycine in 800 mL double-distilled water. When dissolved, adjust volume to 1 L and autoclave.

4. 1M Triethanolamine: Mix 44.5 mL triethanolamine in 200 mL double-distilled water, adjust to pH 8.0 with HCl, then bring to 300 mL volume with double-distilled water before autoclaving.

5. 10X SSC (standard saline citrate): Dissolve 87.65 g NaCl and 44.1 g sodium citrate in 800 mL double-distilled water, adjust to pH 7.0 with 10N NaOH, then bring to 1 L vol. Autoclave a small aliquot to be used for the hybridization buffer. When the solution is used for posthybridization washes, it does not need to be autoclaved (*see* **Note 1**).

6. Deionized formamide: Mix 50 mL of formamide and 5 g of mixed-bed ion-exchange resin (e.g., Bio-Rad AG 501-X8, 20–50 mesh). Stir for 30 min at room temperature, then filter twice through Whatman No.1 filter paper. Store in small aliquots at −20°C.

7. 100X Denhardt's solution: Dissolve 1 g Ficoll, 1 g polyvinylpyrrolidone, and 1 g bovine serum albumin (BSA) (Fraction V) in 50 mL of sterile distilled water and autoclave.

8. Herring sperm DNA: Dissolve the DNA (Type XIV sodium salt) in sterile distilled water at a concentration of 10 mg/mL. If necessary, stir the solution on a magnetic stirrer for 2–4 h at 37–40°C to help the DNA to dissolve. Shear the DNA by passing it several times through a sterile 18-g hypodermic needle. Alternatively sonicate on medium-high power for 5 min. Boil the DNA solution for 10 min and store at −20°C in small aliquots. Just before use, heat the DNA for 5 min in a boiling water bath. Chill it quickly in ice-water.

9. 10% Sodium dodecyl sulfate (SDS): Dissolve 100 g SDS in 900 mL of double-distilled water. Heat to 68°C to assist solubilization. Adjust to pH 7.2 by adding a few drops of HCl. Adjust volume to 1 L. This solution does not need autoclaving, as it is an RNase inhibitor.

10. Hybridization buffer: 50% deionized formamide, 5X SSC, 10% dextran sulfate, 5X Denhardt's solution, 2% SDS, and 100 µg/mL denatured sheared herring sperm DNA. Make this solution fresh before use, and store at 50°C.

11. Phosphate buffered saline (PBS): dissolve in 9 L of distilled water the following: 87.9 g NaCl, 2.72 g KH_2PO_4, 11.35 g Na_2HPO_4 anhydrous. Adjust the pH to 7.1–7.2 with HCl before adjusting the total volume to 10 L. Autoclave before use.

12. 0.2% Triton-X100 (v/v) in autoclaved PBS.
13. Proteinase K (stock solution): Dissolve the proteinase K at 0.5 mg/mL concentration in sterile distilled water. Divide into small aliquots and store at –20°C.
14. Permeabilizing solution: Prepare 0.1M Tris-HCl, pH 8.0, 50 mM EDTA from stock solutions (1M and 0.5M respectively), diluting 1/10 with sterile distilled water. Just before use (**Subheading 3.1., step 2**), to 100 mL of this solution, prewarmed at 37°C, add 200 μL proteinase K stock solution (final concentration 1 μg/mL).
15. 0.1M Glycine in PBS: Dilute 1M stock solution 1/10 with autoclaved PBS.
16. 4% Paraformaldehyde in PBS: Dissolve 4 g of paraformaldehyde in 80 mL of 0.01M PBS with heat, keeping the temperature below 60°C. Stir the slurry until the powder is completely dissolved, and if necessary add few drops of 10N NaOH to clear the solution. Adjust the volume to 100 mL and leave the solution to cool before using. The fixative should be freshly prepared before use.
17. Acetic anhydride.
18. 0.1M Triethanolamine: freshly prepared 1/10 dilution of 1M stock solution in sterile distilled water.
19. Sterile double-distilled water.
20. Digoxigenin-labeled cRNA probe. The probe should be complementary to the target mRNA. The probe should be ethanol precipitated and dissolved in hybridization buffer immediately before use (**Subheading 3.1., step 8**).
21. 5X SSC (*see* **Note 1**).
22. 2X SSC, 0.1% SDS (*see* **Note 1**).
23. 0.1X SSC, 0.1% SDS (*see* **Note 1**).
24. 2X SSC (*see* **Note 1**).
25. RNase stock solution: Dissolve pancreatic RNase (RNase A) at a concentration of 10 mg/mL in 10 mM Tris-HCl, pH 7.5, and 15 mM NaCl. Dispense into aliquots and store at –20°C.
26. 10 μg/mL RNase in 2X SSC: Add 100 μL of RNase stock solution to 100 mL 2X SSC prewarmed at 37°C. Prepare freshly before use.
27. Sections on slides.
28. Buffer 1: 0.1M Tris-HCl, pH 7.5, 0.1M NaCl, 2 mM MgCl$_2$, 3% BSA. In 800 mL double-distilled water dissolve 12.1 g Tris-base and 5.85 g NaCl. Adjust to pH 7.5, then add 0.2 g MgCl$_2$. Adjust the volume to 1 L and add BSA to 3%.
29. Buffer 2: 0.1M Tris-HCl, pH 9.5, 0.1M NaCl, 50 mM MgCl$_2$. In 800 mL double-distilled water dissolve 12.1 g Tris-base and 5.85 g NaCl. Adjust to pH 9.5, then add 4.4 g MgCl$_2$. Adjust the volume to 1 L.
30. Antidigoxigenin antibody/alkaline phosphatase conjugate (Boerhinger, Mannheim, Germany): Immediately before use (**Subheading 3.2., step 3**) dilute to 1/500 with buffer 1.
31. Nitroblue tetrazolium chloride (NBT): Dissolve 35 mg NBT in 277 μL 70% dimethylformamide (DMF). Prepare freshly before use.
32. 5-Bromo-4-chloro-3-indolyl-phosphate (BCIP): Dissolve 17 mg BCIP in 222 μL 100% DMF. Prepare freshly before use.

33. Substrate buffer: In 100 mL buffer 2 dissolve 25 mg levamisole (Sigma, St. Louis, MO), then add 277 µL NBT solution and 222 µL BCIP solution just before use.
34. Stop buffer: 20 mM Tris-HCl, pH 7.5, 5 mM EDTA. In 800 mL double-distilled water dissolve 2.42 g Tris-base, then adjust to pH 7.5 with HCl. Add 10 mL 0.5M EDTA stock solution and adjust volume to 1 L.
35. 5% Pyronin Y: Dissolve 5 g pyronin Y in 100 mL double-distilled water.

3. Methods

All the steps up to hybridization (included) should be carried out in RNase-free conditions. Solutions should be autoclaved or prepared with sterile ingredients using RNase-free equipment. Equipment should be autoclaved, or baked at 250°C for 4 h, as appropriate.

Select the slides, number, and mark them as necessary with pencil (not pen—ink may disappear during the various incubation steps).

3.1. Hybridization

1. Rehydrate the sections by immersion in 0.2% Triton/PBS for 15 min. Wash in PBS twice for 3 min.
2. Carry out the tissue permeabilization by incubating the tissue in permeabilizing solution prewarmed at 37°C, containing 1 µg/mL proteinase K. The normal incubation time is 15–20 min (*see* **Note 2**).
3. Stop the proteinase K activity by immersion in 0.1M glycine in PBS for 5 min.
4. Immerse the sections in 4% paraformaldehyde for 3 min to postfix the target nucleic acid.
5. Rinse the sections briefly in PBS, twice, to remove the paraformaldehyde.
6. Place the slides in a staining jar containing 0.1M triethanolamine and, while stirring, add acetic anhydride to 0.25% (v/v) and incubate for 10 min (*see* **Note 3**).
7. Rinse the slides briefly in double-distilled water and dry them at 37–40°C. This takes approx 10 min.
8. Dissolve the probe in hybridization buffer at 50°C to a final concentration of 2.5 ng/µL.
9. Apply 10 µL of diluted probe per section to the dry slides. The volume of diluted probe can be increased for large sections.
10. Using fine forceps, gently place a siliconized cover slip onto the section to spread the probe solution. If there are any air bubbles, remove them by pressing gently on the cover slip with the forceps.
11. Hybridize the section for 16–20 h (*see* **Note 4**) at suitable temperature in a sealed humid chamber containing 5X SSC. A different hybridization temperature will be needed for various probes according to their T_m (*see* **Note 5**).
12. Following hybridization, remove the cover slip by immersing the slide in 2X SSC, 0.1% SDS. The cover slips will float off after few minutes soaking.
13. Wash the sections in 2X SSC, 0.1% SDS at room temperature, shaking gently, for four changes of 5 min.
14. Wash the slides in 0.1X SSC, 0.1% SDS at the same temperature used for hybridization, shaking gently, for two changes of 10 min (*see* **Note 6**).

15. If using cRNA probes, rinse the sections briefly in 2X SSC, twice, then incubate 10 µg/mL RNase A solution in 2X SSC at 37°C for 15 min (*see* **Note 7**).

16. Rinse briefly in 2X SSC, then PBS before proceeding with the immunohisto-chemistry detection.

3.2. Immunohistochemistry Detection (see Note 8)

1. Block nonspecific binding by immersing the slides in buffer 1 for 10 min at room temperature.

2. Wipe dry the slides around the tissue but keep the tissue wet.

3. Put on the section a drop of antidigoxigenin antisera conjugated to alkaline phosphatase, diluted 1/500 with buffer 1. Incubate 2 h at room temperature (*see* **Notes 8 and 9**).

4. Wash in buffer 1 for three changes of 3 min.

5. Equilibrate the sections in buffer 2 for 10 min at room temperature.

6. Immerse the slides in substrate buffer for 10–30 min (or longer if necessary) at room temperature, covering the dish with aluminum foil to keep the reaction in the dark (*see* **Note 10**).

7. Check the slides under the microscope to assess the development reaction. Stop the reaction or put the slides back in the solution and leave for a longer time as required. The color of the reaction is blue-black.

8. Stop the reaction by immersing in stop buffer for 5 min at room temperature.

9. Counterstain by dipping the slides in 5% pyronin Y solution, for 10–30 s.

10. Rinse well under tap water, approx 5–10 min, until the water is clear.

11. While still wet, mount the slides in aqueous mountant (e.g., Hydromount or similar).

4. Notes

1. There is no need to autoclave the solutions used for posthybridization washes.

2. The incubation time for proteinase K should be titrated for tissue type, as prolonged proteinase digestion could damage the tissue, with a loss of morphology and of target nucleic acid.

3. The treatment with triethanolamine is carried out in order to acetylate the tissue. This prevents electrostatic interaction between the tissue and the probe, as a result of opposite electrostatic charges, thus reducing the background staining.

4. It has been demonstrated that the hybridization reaction reaches an equilibrium after 4–6 h incubation, when a maximum of hybrid has formed. However, incubation is generally carried out overnight for convenience.

5. The hybridization temperature is dependent on the type of probes that have been used. With hybridization buffer containing 50% formamide, it is suggested that the following range of temperature should be tested initially: 42–48°C for cDNA probes; 42–55°C for cRNA probes; 37–40°C for oligonucleotide probes.

6. High background signal can be removed by prolonged washes of the slides at higher stringencies. Increase the number of washes in 0.1X SSC, 0.1% SDS,

also progressively increase the temperature. Take care not to allow the sections to dry out in between any of the washes. Use the same container throughout, changing the solutions quickly.

7. It is essential to remove any trace of SDS from the sections before the incubation with RNase solution, as SDS would inhibit the action of the enzyme. It is desirable to use RNase at this stage of the hybridization procedure if you have been using cRNA probes. RNaseA will degrade the single-stranded cRNA probe that is bound nonspecifically to the section, hence decreasing the background staining. Double-stranded RNA hybrids (cRNA-mRNA) are unaffected by the enzyme.

8. There is a variety of reporter molecules (e.g., alkaline phosphatase, FITC, and so on) conjugated to antibodies against digoxigenin, available from Boehringer, which may be suitable for different applications. There are also detection kits available from the same supplier, which include all reagents needed for the procedure. It is suggested that the reader refers to the catalog from Boehringer for further details.

9. Antibodies conjugated to other reporter molecules may require different dilutions, which are specified by the supplier.

10. The developing time varies considerably according to the type of tissue and the abundance of the target nucleic acid within the cell. In some cases several hours, or overnight, incubation in substrate buffer is necessary to obtain detectable signal. However, it has to be remembered that prolonged incubation also increases the nonspecific background staining.

References

1. Kessler, C. (1991) The digoxigenin anti-digoxigenin (DIG) technology—A survey on the concept and realization of a novel bioanalytical indicator system. *Mol. Cell. Probes* **5**, 161–205.

2. Morris, R. G., Arends, M. J., Bishop, P. E., Sizer, K., Duvall, E., and Bird, C. C. (1990) Sensitivity of digoxigenin and biotin labeled probes for detection of human papillomavirus by *in situ* hybridization. *J. Clin. Pathol.* **43**, 800–805.

3. Furuta, Y., Shinohara, T., Sano, K., Meguro, M., and Nagashima, K. (1990) *In situ* hybridization with digoxigenin-labeled DNA probes for detection of viral genomes. *J. Clin. Pathol.* **43**, 806–809.

4. Giaid, A., Hamid, Q., Adams, C., Springall, D. R., Terenghi, G., and Polak, J. M. (1989) Non-isotopic RNA probes. Comparison between different labels and detection systems. *Histochemistry* **93**, 191–196.

5. Dirks, R. W., van Gijlswijk, R. P. M., Tullis, R. H., Smit, A. B., van Minnen, J., van der Ploeg, M., and Raap, A. K. (1990) Simultaneous detection of different mRNA sequences coding for neuropeptide hormones by double *in situ* hybridization using FITC- and biotin-labeled oligonucleotides. *J. Histochem. Cytochem.* **38**, 467–473.

6. Herrington, C. S., Burns, J., Graham, A. K., Bhatt, B., and McGee, J. O. D. (1989) Interphase cytogenetics using biotin and digoxigenin labeled probes. II: Simultaneous detection of two nucleic acid species in individual nuclei. *J. Clin. Pathol.* **42**, 601–606.

7. Trask, B. J., Massa, H., Kenwrick, S., and Gitschier, J. (1991) Mapping of human chromosome Xq28 by two colour fluorescence *in situ* hybridization of DNA sequences to interphase cell nuclei. *Am. J. Hum. Genet.* **48**, 1–15.

22

One-Tube RT-PCR with Sequence-Specific Primers

Ulrich Pfeffer

1. Introduction

From the very beginning of PCR technology *(1–3)*, it was clear that the power of amplification could be used for the study of mRNA expression *(4,5)*. This technique is now widely used and many different protocols have been developed using either viral reverse transcriptases (RT) or exploiting the reverse transcriptase activity inherent in many thermostable DNA polymerases *(6)*. The main problem related to this technique when compared to other methods of RNA analysis is the primer design, which remains mainly empirical. The normal approach to reduce nonspecific reactions is the use of nested primers which, however, further complicates the reaction scheme. Another possibility is the use of a sequence-specific RT primer different from the PCR primers which, however, must be removed before PCR amplification.

We have developed a simple protocol that makes use of a short sequence-specific RT primer (normally a dodecamer; ref. *7*) that must not be removed because it does not bind to the target at the higher amplification temperatures. The use of sequence-specific RT primers presents several advantages:

1. The specificity of RT-PCR is increased because only those RNAs that contain a sequence complementary to the RT-primer are reverse transcribed.
2. The dodecamer primer improves the analysis of low abundance RNA molecules with long 3' untranslated regions, since reverse transcription starts close to the region to be amplified.
3. In general, shorter cDNAs are required and thus more suitable template is produced,
4. Regions with secondary structures where reverse transcriptases show reduced processivity can be avoided when designing the primers.

The major drawback of this approach is the difficulty in carrying out many amplifications with the same cDNA. In some cases it may be desirable to

From: *Methods in Molecular Biology, Vol. 86: RNA Isolation and Characterization Protocols*
Edited by: R. Rapley and D. L. Manning © Humana Press Inc., Totowa, NJ

reverse transcribe an RNA sample once and to store the cDNA for any future amplification. In this case, oligo-dT or random hexamers should be used. These primers are perfectly compatible with the system described here and may be substituted for the specific primer. However, the specificity and yield of the reaction often drop drastically, especially if low abundance messenger RNAs with long 3' untranslated regions are analyzed. In most cases, the use of short sequence specific primers yields more and cleaner amplification products. More importantly, the design of primers for new sequences results in successful first attempts, without lengthy optimizations.

The protocol presented here has been developed using standard RT PCR protocols *(4,5)*, as shown in **Figs. 1** and **2**. Much effort has been devoted to the simplification of the procedure in the attempt to reduce sample handling and the risk of contamination and to increase the sample-to-sample accuracy.

RT primer dodecamers are selected closely downstream to the PCR antisense primer and may also be contiguous to it. T_m values of RT primers should be between 25 and 35°C. If possible, the 3' nucleotides should be G or C in order to assure stable base pairing at the 3' end and not show dimerization capabilities. In the region between the RT primer and the PCR sense primer no particularly GC-rich regions, which could interfere with the processivity of the reverse transcriptase, should occur (*see also* **ref. 8** and **Note 1**). Several dodecamer primers may be used in the same reverse transcription reaction provided that they are not complementary to each other. In this way, a sample mRNA and an internal standard mRNA, such as β-actin or glycerol-3-phosphate dehydrogenase mRNAs, can be reverse transcribed in the same tube. The subsequent amplification of the two cDNA is in general not suitable since nonspecific priming and competition may occur. PCR primers are selected, if possible, to anneal to different exons in order to avoid amplification from traces of genomic DNA eventually contaminating RNA preparations. The final amplification product should be of 200–1000 bp in length, a range in which amplification works best and in which the resolution of agarose gels is the highest. **Figure 2** shows an example of the glucocorticoid receptor mRNA *(9,10)* and some features that must be considered when designing primers.

Reverse transcription is carried out in two steps at 25 and 42°C. At the lower temperature the primer anneals to the complementary sequence in the mRNA and is slowly extended by the reverse transcriptase. The extended primer remains annealed to the template also at the higher temperature optimal for the enzyme, and allows efficient extension. The use of dodecamer RT primers is compatible with both Moloney Murine Leukemia Virus (MuLV) and Avian Myeloblastosis Virus (AMV) reverse transcriptases but not with protocols that exploit the reverse transcriptase activity of thermostable DNA polymerases such as that isolated from *Thermus thermophilus* (Tth; **ref. 6**), since these

```
   1 TTTTTAGAAAAAAAAAAATATATTTCCCTCCTGCTCCTTCTGCGTTCACAA
  51 GCTAAGTTGTTTATCTCGGCTGCGGCGGGAACTGCGGACGGTGGCGGGCG      G/C rich
 101 AGCGGCTCCTCTGCCAGAG/TTGATATTCACTGATGGACTCCAAAGAATCA      exon 2
 151 TTAACTCCTGGTAGAGAAGAAAACCCCAGCAGTGTGCTTGCTCAGGAGAG
 201 GGGAGATGTGATGGACTTCTATAAAACCCTAAGAGGAGGAGCTACTGTGA
 251 AGGTTTCTGCGTCTTCACCCTCACTGGCTGTCGCTTCTCAATCAGACTCC
 301 AAGCAGCAGCCAGATCTGTCCAAAGCAGTTTCACTCTCAATGGGACTGTATA
 351 GCAGCAGCCAGATCTGTCCAAAGCAGTTTCACTCTCAATGGGACTGTATA
 401 TGGGAGAGACAGAAACAAAAGTGATGGGAAATGACCTGGGATTCCCACAG
 451 CAGGGCCAAATCAGCCTTTCCTCGGGGGAAACAGACTTAAAGCTTTTGGA
 501 AGAAAGCATTGCAAACCTCAATAGGTCGACCAGTGTTCCAGAGAACCCCA
 551 AGAGTTCAGCATCCACTGCTGTGTCTGCTGCCCCCACAGAGAAGGAGTTT
 601 CCAAAAACTCACTCTGATGTATCTTCAGAACAGCAACATTTGAAGGGCCA
 651 GACTGGCACCAACGGTGGCAATGTGAAATTGTATACCACAGACCAAAGCA
 701 CCTTTGACATTTTGCAGGATTTGGAGTTTTCTTCTGGGTCCCCAGGTAAA
 751 GAGACGAATGAGAGTCCTTGGAGATCAGACCTGTTGATAGATGAAAACTG
 801 TTTGCTTTCTCCTCTGGCGGGAGAAGACGATTCATTCCTTTTGGAAGGAA
 851 ACTCGAATGAGGACTGCAAGCCTCTCATTTTACCGGACACTAAACCCAAA
 901 ATTAAGGATAATGGAGATCTGGTTTTGTCAAGCCCCAGTAATGTAACACT
 951 GCCCCAAGTGAAAACAGAAAAAGAAGATTTCATCGAACTCTGCACCCCTG
1001 GGGTAATTAAGCAAGAGAAACTGGGCACAGTTTACTGTCAGGCAAGCTTT
1051 CCTGGAGCAAATATAATTGGTAATAAAATGTCTGCCATTTCTGTTCATGG
1101 TGTGAGTACCTCTGGAGGACAGATGTACCACTATGACATGAATACAGCAT
1151 CCCTTTCTCAACAGCAGGATCAGGAGCCTATTTTTAATGTCATTCCACCA
1201 ATTCCCGTTGGTTCCGAAAATTGGAATAGGTGCCAAGGATCTGGAGATGA
1251 CAACTTGACTTCTCTGGGGACTCTGAACTTCCCTGGTCGAACAGTTTTTT      sense primer
1301 CTAATGGCTATTCAAG/CCCCAGCATGAGACCAGATGTAAGCTCTCCTCCA      exon 3
1351 TCCAGCTCCTCAACAGCAACAACAGGACCACCTCCCAAACTCTGCCTGGT
1401 GTGCTCTGATGAAGCTTCAGGATGTCATTATGGAGTCTTAACTTGTGGAA
1451 GCTGTAAAGTTTTCTTCAAAAGAGCAGTGGAAG/GACAGCACAATTACCTA      exon 4
1501 TGTGCTGGAAGGAATGATTGCATCATCGATAAAAATTCGAAGAAAAAACTG
1551 CCCAGCATGCCGCTATCGAAAATGTCTTCAGGCTGGAATGAACCTGGAAG
1601 /CTCGAAAAACAAAGAAAAAAAATAAAAGGAATTCAGCAGGCCACTACAGGA      exon 5
1651 GTCTCACAAGAAACCTCTGAAAATCCTGGTAACAAAACAATAGTTCCTGC
1701 AACGTTACCACAACTCACCCCTACCCTGGTGTCACTGTTGGAGGTTATTG
1751 AACCTGAAGTGTTATATGCAGGATATGATAGCTCTGTTCCAGACTCAACT
1801 TGGAGGATCATGACTACGCTCAACATGTTAGGAGGGCGGCAAGTGATTGC
1851 AGCAGTGAAATGGGCAAAGGCAATACCAG/GTTTCAGGAACTTACACCTGG      exon 6
1901 ATGACCAAATGACCCTACTGCAGTACTCCTGGATGTTTCTTATGGCATTT
1951 GCTCTGGGGTGGAGATCATATAGACAATCAAGTGCAAACCTGCTGTGTTT
2001 TGCTCCTGATCTGATTATTAATGAG/CAGAGAATGACTCTACCCTGCATGT      exon 7
2051 ACGACCAATGTAAACACATGCTGTATGTTTCCTCTGAGTTACACAGGCTT
2101 CAGGTATCTTATGAAGAGTATCTCTGTATGAAAACCTTACTTGCTTCTCTC
2151 TTCAG/TTCCTAAGGACGGTCTGAAGAGCCAAGAGCTATTTGATGAAATTA      exon 8
2201 GAATGACCTACATCAAAGAGCTAGGAAAAGCCATTGTCAAGAGGGAAGga      antisense primer
2251 aactccagccAGAACTGGCAGCGGTTTTATCAACTGACAAAACTCTTGGA      rt-primer
2301 TTCTATGCATGAAG/TGGTTGAAAATCTCCTTAACTATTGCTTCCAAACAT      exon 9
2351 TTTTTGGATAAGACCATGAGTATTGAATTCCCCGAGATGTTAGCTGAAATC
2401 ATCACCAATCAGATACCAAAATATTCAAATGGAAATATCAAAAAACTTCT
2451 GTTTCATCAAAAGTGACTGCCTTAATAAGAATGGTTGCCTTAAAGAAAGT
2501 CGAATTAATAGCTTTTATT............................      extended 3'UT-
                                    ..//..
4701 ..........AATGCTTTTTGAATGTTATTTGTTATTTTCAGTATTTTGG
4751 AGAAATTATTTAATAAAAAAACAATCATTTGCTTTTTG
```

Fig. 1. RT-PCR primers for the glucocorticoid receptor; for the primer design the following criteria have been considered: PCR primers are on different exons, i.e., the G/C-rich 5' region has been avoided; the RT primer is close to the antisense primer thus avoiding reverse transcription of the long 3' untranslated region (>2200 nt); and Primers are neither self- nor cross-complementary (sequence from **ref. 9**, genomic organisation from **ref. 10**; antisense and RT primers are indicated on the upper strand, the actual primer sequences are the inverted complements of the sequences shown).

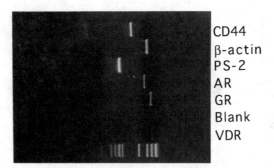

CD44
β-actin
PS-2
AR
GR
Blank
VDR

Fig. 2. RT-PCR using dodecamer RT-primers total RNAs isolated from MDA-MB 231 or MCF-7 human mammary carcinoma cells were reverse-transcribed using sequence-specific dodecamer primers and amplified as described in the text (35 cycles). Half of each of the different cDNAs were used for amplification, except for β-actin in which one tenth has been used. CD44 = cell determinant 44, hyalurone receptor; PS-2, breast cancer marker; AR, androgen receptor; GR, glucocorticoid receptor; blank, reaction without RNA using GR primers; VDR, vitamin D receptor. Φx-174 HaeIII digest has been applied as a length marker (right lane; marker band lengths from top: 1353, 1078, 872, 603, 310, 281/271, 234, 194, 118, 72 bp).

enzymes perform the reverse transcription at temperatures where the short sequence primers do not anneal to the template.

The cDNA obtained is either used as a whole or divided for multiple amplifications and/or amplification of the internal standard. PCR primers and Taq polymerase are added, cDNA is denatured and then amplified. **Figure 1** shows a series of different analyses carried out following this protocol using primer sets designed in our laboratory.

When performing semiquantitative analyses, particular care must be taken in order to assure that none of the components of the system is exhaused during the reaction leading to saturation. Different amounts of sample RNA are analyzed by RT-PCR using primers for a sample RNA and an internal standard messenger RNA. The correlation between the amount of starting RNA and the amount of amplification product obtained must be linear for both the sample and the standard of the gels or by inclusion of a radioactive tracer nucleotide and scintillation counting after acid precipitation.

2. Materials

1. Total RNA may be prepared by any of the common procedures provided that the preparation results are essentially devoid of proteins. In our hands, the method described by Chomczynski and Sacchi *(11)* and the commercial formulations of it work best.
2. RNA should be redissolved in H_2O or in TE, no detergents should be added (*see* **Note 3**).

3. Nucleotide stocks are titrated with NaOH to pH 7.0 or purchased as premade buffered solutions (Pharmacia, Brussels, Belgium).

4. Oligonucleotides should be purified at least by gel filtration and the success of synthesis should be controlled. FPLC purified oligonucleotide primers used here were obtained from TIB-Molbiol, Genoa, Italy.

5. RNase inhibitor and AMV reverse transcriptase may be obtained from Boehringer, Mannheim and Taq-polymerase from AGS, Heidelberg, Germany. Enzymes provided by other suppliers perform equally well (*see* **Note 4**).

6. PCR-buffer (1X): 10 mM Tris-HCl pH 8.3, 50 mM KCl, 2 mM MgCl$_2$, 0.01% Tween-20 (*see* **Note 5**).

7. Pre RT-PCR and post RT-PCR components and equipment should be physically separated, if possible in a one-way pursuit. The use of filter pipet tips is recommended (*see* **Note 6**).

3. Methods
3.1. Primer Design

Select PCR sense and antisense primers using gene analysis software (*see* **Note 1**). Make sure that the PCR primers do not show self- or cross-homology, partial homology with other regions of the sequence analyzed or with sequences present in the databank. T$_m$ values should be between 75–85°C, thus allowing annealing temperatures of 65°C. Then select a dodecamer RT primer closely downstream to the PCR antisense primer. This primer should not be complementary to the PCR primers or to the RT primer of the internal standard that eventually will be analyzed in parallel. Avoid self complementary primers; a restriction site present in the dodecamer is likely to inactivate it. T$_m$ values of RT primers should be between 25–35°C. Unless specific regions of the messenger are to be analyzed, you can freely choose your primers, but make sure that no GC-rich regions are contained between the RT-primer and the PCR sense primer (complete analyzed region). Choose primers on different exons. If exon/intron borders are not known choose distant PCR primers *(12)* and check for amplification from DNA. For some applications, for example RT-PCR from RNA extracted from paraffin-embedded tissue samples, primers must be close to each other in order to allow amplification from partially degraded RNA *(13)*. After synthesis, primers must be purified by gel filtration in order to remove protecting groups and salts. Normally, no further purification occurs but the actual success of synthesis should be monitored by an analytical cromatography run. If products of precocious chain termination are present, the primers should be purified.

3.2. Reverse Transcription (see Notes 7 and 8)

1. Denature 5 µL RNA solution (1 µg total RNA or 0.01 µg polyA$^+$-RNA) at 65°C for 10 min (*see* **Note 9**).

2. In the meantime prepare the RT mix for the number of samples to be analyzed:
   1X PCR buffer
   1 mM dNTP
   25 pMol RT-primer
   20 U RNasin
   2 U AMV reverse transcriptase
   H_2O to 15 µl (per sample)
   20 µL final volume per sample (after addition of RNA)
3. Keep the denatured RNA on ice and add 15 µL RT mix to each sample, vortex and spin in a microfuge to collect the sample at the bottom of the tube (*see* **Note 10**).
4. Incubate at 25°C, 10 min and at 42°C, 45 min (*see* **Note 11**), keep on ice until further processing (*see* **Notes 12** and **13**).

3.3. Polymerase Chain Reaction (see Notes 7 and 8)

1. Prepare the PCR-mix for the number of samples to be analyzed:
   1X PCR buffer
   0.2 mM dNTP (*see* **Note 14**)
   25 pMol sense PCR primer
   25 pMol antisense PCR primer
   2.5 U Taq polymerase
   H_2O to final volume
   100 µL final volume per sample (after addition of cDNA; *see* **Note 15**)
2. Keep cDNA samples on ice and add the PCR mix to a final volume of 100 µL, vortex, spin to collect the sample at the bottom of the tube, add two drops of mineral oil.
3. Cycle with the following settings (combined protocol):
   A. Denaturation (*see* **Note 16**)
   94°C, 2 min, 1 cycle
   B. Amplification
   94°C, 30 min
   55–65°C, 30 min (*see* **Note 17**)
   72°C, 30–60 min (*see* **Note 18**)
   25–35 cycles (*see* **Note 19**)
   C. Final extension and storage
   72°C, 7 min
   4°C, ∞
   1 cycle
4. Remove 10 µL of each sample and analyze on a 1–1.5% agarose gel containing 0.4 µg/mL ethidium bromide (care: potent carcinogen!; *see* **Note 20**)

4. Notes

1. We use Primer Detective (Clontech, Palo Alto, CA) and Oligo (Medprobe, Oslo, Norway) software for primer design, but we have not analyzed others that may be equally suited. Not all programs (as Primer Detective) allow the analysis of

single primers necessary for the design of the dodecamer and most do not analyze structural features of the region between the primers. In our experience, it is sufficient to exclude regions of high G/C content and this can be carried out manually.

2. When performing semiquantitative analyses, an internal standard messenger must be amplified in parallel. The RT reaction can be performed together with the sample but in general, coamplification will not be successful. The standard messenger is normally expressed at higher levels than the sample messenger and thus competes for system components and the presence of more PCR primers (four instead of two) increases the possibility of nonspecific amplification. However, the major source of error resides in the quantity of RNA starting concentration and in RNA degradation during handling and in pipeting the relatively small amounts of the RT components. These errors are adequately controlled when performing combined RT reactions but separated PCR reactions.

3. The analysis may be performed on either total or poly-A$^+$mRNA. The quantity necessary to obtain a UV-visible band on the gel depends on the level of expression of the given sample messenger RNA. When starting with 1 µg total RNA or 0.01 µg poly-A$^+$mRNA, using half of the cDNA for amplification and performing 35 cycles, low abundance messengers should yield a product. Overdrying of RNA pellets results in incomplete redissolution. DNA contaminations are to be removed with RNase-free DNAse I if the PCR primers are on the same exon. All solutions for RNA and for reverse transcription should be RNase-free.

4. We did not encounter problems with the quality of enzymes purchased from various sources but noted that the enzymes are the most expensive part of the assay.

5. The PCR buffer indicated is one of many possible ones. In most cases the buffer provided with the Taq polymerase is suitable. The MgCl$_2$ concentration in both reverse transcription and amplification is critical and should be optimized for each new case. In our hands, 2 mM is a good starting concentration. If under the conditions described no amplification product can be obtained, the simple variation of the Mg^{2+} concentration will not improve the situation, rather its variation may result in an improvement of already working amplification conditions.

6. If contamination occurs, all aliquots already used should be discarded, the environment should be cleaned and equipment should be autoclaved where possible (there are autoclavable pipets!). General amplification failure (no product) is normally due to RNase contaminations derived from the hands of the operator or from RNases used in the same lab (e.g., plasmid preps!). Run positive controls of RNA preparations from cells that express the gene analyzed. If amplification products of unexpected length occur, consider the possibility of alternatively spliced messenger RNAs. This can be verified by hybridization of the blotted amplification product to a specific probe.

7. All components of the RT mix and of the PCR mix, except the enzymes and the RNase inhibitor can be assembled in large master mixes, aliquoted, and stored at −20°C for extended periods. This enhances the sample-to-sample reproducibility and diminishes the pipeting steps to be carried out on the single sample, also reducing the risk of contaminations.

8. The major pitfall of RT-PCR methods are contaminations with PCR products of previous analyses and with plasmids containing related sequences. Always perform blank analyses in which the RNA solution has been substituted with plain water. Post-PCR processing and plasmid preparations should not be carried out in the same room as the pre-PCR steps, a set of pipets and of other equipment should be reserved for pre-PCR operations.

9. Denaturation of RNA may also be performed for 5 min at 80°C, the use of formamide is not recommended. After denaturation the RNA is to be kept on ice to avoid renaturation and further processing should be carried out soon.

10. Although mixing of the reaction components is important, we have observed that vortexing (and subsequent centrifugation) can be omitted. Do not mix by repeated pipeting, this is one of the major sources of pipet contamination!

11. The annealing temperature of the dodecamer primer is 25°C, 42°C is the temperature optimum for AMV reverse transcriptase. Other transcriptases such as MuLV have an optimum of 37°C but they may also be used at 42°C. The processivity and the capability to read through "difficult" sequences of AMV is somewhat higher.

12. If you wish to interrupt the procedure, you can do that at the end of the cDNA synthesis. cDNA is stored at 4°C for prolonged periods, storage at −20°C or lower is not recommended.

13. There is no need to destroy reverse transcriptase at the end of the cDNA synthesis.

14. Take into account the carryover from the reverse transcription reaction when setting up the PCR mix. If you use 10 mL of the cDNA solution, the dNTPs carried over account for half of the amount needed in the PCR reaction.

15. The amount of the cDNA to be used for amplification depends on various considerations:
    i. expression level;
    ii. eventual parallel amplification of an internal standard when the reverse transcription has been carried out in the presence of the appropriate primers;
    iii. ease of performance: only if the sample messenger RNA is to be analyzed in a large number of samples it may prove easier just to add the PCR mix to the whole cDNA preparation; and
    iv. eventual storage of a part of the cDNA for further analysis.

16. Denaturation prior to PCR is carried out in the thermal cycler in the presence of the Taq polymerase. The samples should be introduced into the hot cycler. Two minutes is sufficient for the denaturation of the relatively short cDNAs, any longer results in a loss of enzyme activity. If the thermal cycler does not allow this type of programming, denaturation may be performed separately.

17. Annealing should be performed at 5–10°C below the thermodynamic T_m value of the primer with the lowest T_m. In practice it may be useful to use long primers (25-mers) that can be used at higher annealing temperatures.

18. The Taq polymerase should incorporate at least 1500 nt per minute so that 30 s of extension are enough in most cases. Longer extension periods often described in the literature lead to premature exhaustion of the enzyme and reduce the number of cycles that can be performed without adding fresh enzyme.

19. In order to enhance the sensitivity of the system, more cycles may be performed. After 40 cycles the enzyme is expected to be exhausted and fresh enzyme should be added.
20. When master mixes are prepared a large number of samples within the limit of capacity of the thermal cycler can easily be handled in parallel. Results are obtained within a maximum of 5 h with less than 1 h hands-on time.

Acknowledgments

The author thanks Giorgio Vidali for continuous support. Grants from AIRC, CNR, and Ministero della Sanità are acknowledged.

References

1. Saiki, R., Scharf, S., Faloona, F., Mullis, K. B., Horn, G. T., Erlich, H. A., and Arnheim, N. (1985) Enzymatic amplification of β-globin genomic sequences and restriction site analysis for diagnosis of sickle cell anemia. *Science* **230**, 1350–1354.
2. Mullis, K. B., and Faloona, F. A. (1987) Specific synthesis of DNA *in vitro* via a polymerase-catalysed chain reaction. *Methods Enzymol.* **155**, 335–350.
3. Saiki, R. K., Gelfand, D. H., Stoffel, S., Scharf, S. J., Higuchi, R., Horn, G. T., Mullis, K. B., and Erlich, H. A. (1988) Primer-directed enzymatic amplification of DNA with a thermostable DNA polymerase. *Science* **239**, 487–491.
4. Kawasaki, E. S., Clark, S. S., Coyne, M. Y., Smith, S. D., Champlin, R., Witte, O. N., and McCormick, F. P. (1987) Diagnosis of chronic myeloid and acute lymphocytic leukemias by detection of leukemia-specific mRNA sequences amplified *in vitro*. *Proc. Natl. Acad. Sci. USA* **85**, 5698–5702.
5. Rappolee, D. A., Mark, D., Banda, M. J., and Werb, Z. (1988) Wound macrophages express TGF-α and other growth factors *in vivo:* analysis by mRNA phenotyping. *Science* **241**, 708–712.
6. Myers, T. W. and Gelfand, D. H. (1991) Reverse transcription and DNA amplification by a *Thermus thermophilus* DNA polymerase. *Biochem* **30**, 7661–7666.
7. Pfeffer, U., Fecarotta, E., and Vidali, G. (1995) Efficient one-tube RT-PCR amplification of rare transcripts using short sequence specific reverse transcription primers. *BioTechniques* **18**, 204–206.
8. Pallansch, L., Beswick, H., Talian, J., and Zelenka, P. (1990) Use of an RNA folding algorithm to choose regions for amplification by the polymerase chain reaction. *Anal. Biochem.* **185**, 57–62.
9. Hollenberg, S. M., Weinberger, C., Ong, E. S., Cerelli, G., Oro, A., Lebo, R., Thompson, E. B., Rosenfeld, M. G., and Evans, R. M. (1985) Primary structure and expression of a functional human glucocorticoid receptor cDNA. *Nature* **318**, 635–641.
10. Encio, I. J., and Detera-Wadleigh, S. D. (1991) The genomic structure of the human glucocorticoid receptor. *J. Biol. Chem.* **266**, 7182–7188.
11. Chomczynski, P. and Sacchi, N. (1987) Single-step method of RNA isolation by acid guanidinium thiocyanate-phenol-chloroform extraction. *Ann. Biochem.* **162**, 156–159.
12. Hawkins, J. D. (1988) A survey on intron and exon lengths. *Nucl. Acids Res.* **16**, 9893–9908.
13. Mies, C. (1994) A simple, rapid method for isolating RNA from paraffin-embedded tissues for reverse transcription-polymerae chain reaction (RT-PCR). *J. Histochem. Cytochem.* **42**, 811–813.

23

Identification of Differentially Expressed Genes by Nonradioactive Differential Display of Messenger RNA

Thomas C. G. Bosch and Jan U. Lohmann

1. Introduction

Changes in cell behavior are driven by changes in gene expression. Thus, in order to understand the mechanisms regulating cell behavior, one has to identify and characterize differentially expressed genes. Standard methods currently used to isolate differentially expressed genes include subtractive hybridization *(1,2)*, differential hybridization *(3)*, and single-cell polymerase chain reaction (PCR) *(4)*. Differential display of mRNA by PCR (DD-PCR) is a new and powerful procedure for quantitative detection of differentially expressed genes *(5,6)*. Advantages over alternative approaches include: quantitative identification of differences in gene expression between different cell fractions; simultaneous detection of both upregulation and downregulation of genes; requirement of only small amounts of messenger RNA; and drastically reduced time of analysis. Since on average there are about 15,000 individual mRNA species present in any individual cell, some steps of selection have to be taken before the transcript population can be displayed simultaneously. Therefore, in the DD-PCR procedure mRNA from different cell fractions are reverse transcribed (RT) using an oligo-dT-NN anchor primer. Since this 3' primer will hybridize only with transcripts carrying the corresponding two NN-bases in front of the poly(A) tail, only $1/12$ (i.e., about 1200) of all transcripts are reverse transcribed into cDNA. Accordingly, for reverse transcription of all messages 12 different 3' primer have to be used independently. PCR of the resulting cDNA is then carried out using radiolabeled dNTPs, the oligo-dT-NN anchor primer as 3' primer and a short arbitrary oligonucleotide as 5' primer. By using Homers as 5' primers, about 150–300

From: *Methods in Molecular Biology, Vol. 86: RNA Isolation and Characterization Protocols*
Edited by: R. Rapley and D. L. Manning © Humana Press Inc., Totowa, NJ

products will be obtained in a PCR using 1200 different cDNAs as templates. Thus, for quantitative screening of the transcript pool of a particular cell, 20 arbitrary 5' primers have to be used in 240 independent PCR reactions. After electrophoretic separation of the resulting fragments on a polyacrylamide gel, differential gene expression is visualized by autoradiography. Differentially expressed cDNA species can be recovered from the gel using the autoradiogram for band localization. DD-PCR has previously been employed to identify differentially expressed genes in preimplantation mouse embryo *(7)*, in human endothelial cells treated with fibroblast growth factor *(8)*, in normal and tumor cells *(9)*, and in *Salmonella* treated with peroxide *(10)*. A recent review of the DD-PCR method can be found in McClelland et al. *(6)*.

Although straightforward, the original procedure suffers from a number of drawbacks—poor reproducibility and false positives due to the technical difficulty in recovering a unique radiolabeled DNA species from the polyacrylamide gel *(8,11,12)*. Since success rates in differential display are likely to be increased considerably by directly staining the cDNA populations in the polyacrylamide gel, here we describe a simple and nonradioactive method of differential display of cDNA that allows both rapid screening of a large number of samples for differentially expressed genes and efficient isolation of unique cDNA species (REN-Display, ref. *13*).

mRNA is isolated from two or more cell populations, RT and PCR amplified using one 3' anchored oligo-dT-NN primer and one 5' Homer arbitrary primer. The PCR products are resolved in adjacent lanes on horizontal polyacrylamide gels under denaturing conditions. cDNA bands are visualized by silver staining. Differentially expressed mRNA species are identified by comparing the pattern of silver-stained cDNAs in adjacent lanes. To detect false positives, 2–3 identical PCR reactions are carried out in parallel. Differentially expressed mRNA species can then be directly isolated from the gel and further characterized. Due to the visibility of the cDNA in the original gel, the probability of recovering a mRNA species of interest is increased considerably compared to the standard radioactive procedure. However, even single silver-stained bands contain usually more than one transcript species. Therefore, cloning of the PCR products must be followed by analyzing several (up to 10) independent transformants.

The outline of the nonradioactive differential display method is shown in **Fig. 1**. The REN technique is particularly valuable when applied to complex biological systems and developmental and physiological states, since it requires little amounts of biological material, allows rapid (72 h) investigations of a large number of samples, and can be carried out even in routine diagnostic laboratories.

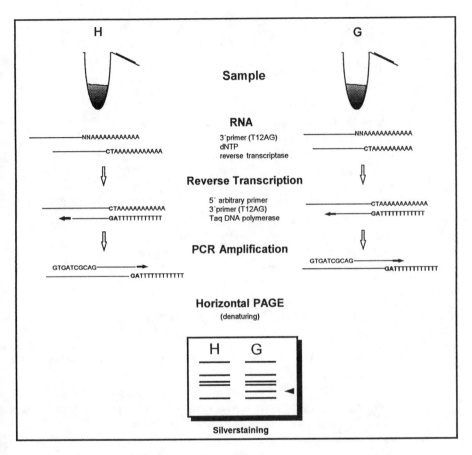

Fig. 1. Outline of the REN display procedure.

2. Materials

2.1. mRNA Preparation and Reverse Transcription

All solutions are prepared with water that has been treated with 0.1% diethyl pyrocarbonate.

1. QuickPrep mRNA Purification Kit (Pharmacia Biotech, Freibrug, Germany).
2. First-strand cDNA Synthesis Kit (Pharmacia Biotech).
3. 3' tailing primers (e.g., 5'-$T_{(12)}$AG-3').

2.2. PCR

1. Thermocycler.
2. *Taq* polymerase.
3. Nucleotide triphosphates.

4. Arbitrary 10-mer oligonucleotides (Operon, Alameda, CA).
5. Sequencing stop solution: 95% formamide, 20 mM Na$_2$EDTA, 0.05% (w/v) bromphenol blue, 0.05% (w/v) xylene cyanol.

2.3. Horizontal Gel Electrophoresis

1. CleanGel48S (ETC Electrophorese Technik, Kirchentellinsfurt, Germany).
2. DNA Disc Buffer system (ETC Electrophorese Technik).
3. Multiphor II Electrophoresis Unit (PharmaciaBiotech).
4. Gelpool/Paperpool (PharmaciaBiotech).
5. Silver staining solutions *(14)*:
   All solutions are prepared in a final volume of 250 mL:
   a. Fixing solution: 10% acetic acid (v/v).
   b. Staining solution: 0.1% AgNO$_3$ (w/v) + 250 μL 37% formaldehyde.
   c. Developing solution: ice-cold 2.5% Na$_2$CO$_3$ + 250 μL 37% formalde-hyde + 250 μL 2% Na-thiosulfate solution.
   d. Stop solution: 2% glycine + 0.5% EDTA-disodium.
   e. Impregnation: 5% glycerol (v/v).

3. Methods

3.1. mRNA Preparation and Reverse Transcription (see Notes 1 and 2)

1. Isolate mRNA or total RNA from about 5×10^5 cells using, for example, the Pharmacia QuickPep mRNA Purification Kit. Final volume: 20 μL.
2. Calculate concentration of RNA by removing an aliquot (2 μL) of RNA suspension and determining the A$_{260}$ in 100 μL water. Yield should be approx 1 μg mRNA (*see* Chapter 12). The samples can be stored at –20°C at this point.
3. For reverse transcription use 250 ng of rnRNA or 3 μg of total RNA and First-Strand cDNA Synthesis Kit from Pharmacia Biotech.
4. Mix 5 μL bulk reaction mix, 1 μL DDT, 1.25 μL of 25 μM stock of tailing primer (e.g., 5'–T$_{(12)}$AG–3 ').
5. Heat 7.75 μL of RNA solution (250 ng of mRNA or 3 μg of total RNA) at 65°C for 10 min. Thereafter place on ice.
6. Add 7.75 μL of mRNA solution to reaction mix. Final volume: 15 μL.
7. Reaction time is 1 h at 37°C.
8. Stop reaction by heating for 5 min at 90°C.

3.2. PCR

1. Dilute the resulting cDNA mixture 1:50 with water. Use 3 μL of a dilution as template for a 10 μL PCR reaction (concentrations are of stock solutions) (*see* **Notes 3** and **4**).
2. Mix 3 μL cDNA (1:50), 1 μL 10X PCR buffer, 1 μL random 10-mer primer (5 μM), 1 μL tailing primer (25 μM), 2 μL dNTP (100 μM), 0.1 μM *Taq* (0.5 U), 1.9 μL H$_2$O.

2. Cover reaction with 30 µL mineral oil.
3. PCR parameters: initial denaturing step for 5 min at 94°C, followed by 43 cycles with cycle times of 30 s at 94°C, 60 s at 42°C, and 30 s at 72°C.
4. Stop reaction by adding 7 µL sequencing stop solution to 10 µL reaction mix. The samples can be stored at –20°C at this point.

3.3. Horizontal Polyacrylamide Gel Electrophoresis

1. Hydrate CleanGels (ETC) in 40 mL gel buffer containing $7M$ urea for at least 1 h on a shaker using Gelpool.
2. Cut two pieces of electrode wicks lengthwise in half and soak them in 40 mL electrode buffer using Paperpool.
3. Apply 2 mL kerosin oil to center of Multiphor II cooling plate.
4. Remove excess buffer from the gel using filter paper.
5. Lay hydrated gel on the plate.
6. Lay electrode wicks on hydrated gel.
7. Boil sample for 4 min. Chill samples on ice before loading.
8. Load 8 µL of sample solution in wells. Store remaining solution at –20°C.
9. Running conditions: Start at 400 V_{max} and 11 mA_{max} for 30 min at room temperature (22°C). Thereafter run at 18 mA_{max} for 2 h.
10. Stop electrophoresis when Xylene cyanol band reaches the anode wicks.

3.4. Visualization of PCR Products by Silver Staining (see Notes 5 and 6)

1. Fix for 30 min in 250 mL fixation solution.
2. Wash 3 × 2 min in 250 mL H_2O.
3. Stain for 30 min in 250 mL staining solution.
4. Rinse gel for 30 s in H_2O.
5. Develop for 5 to 15 min in 250 mL ice-cold developing solution.
6. Stop reaction by incubating for 10 min in 250 mL stop solution.
7. Impregnate by incubating for 10 min in 250 mL impregnation solution.

A typical result showing the differential pattern of silver-stained cDNAs from head and gastric tissue of the freshwater polyp *Hydra* is shown in **Fig. 2**. Poly(A)$^+$ RNA was isolated and reverse transcribed using the 3' anchor primer $T_{(12)}AG$. PCR amplification of the reverse transcription reaction was carried out in duplicates using the 3' anchor primer and an arbitrary Homer 5' primer (5'-GGGTAACGCC-3'). Two differentially expressed transcripts can be detected by this primer (arrowheads in **Fig. 2**).

3.5. Elution of Candidate Bands and Reamplification (see Note 7)

1. Transfer silver-stained cDNA directly from the gel into a PCR tube using two sterile pipet tips.
2. Add 6 µL water to the isolated polyacrylamide slice.
3. Smash the gel slice and vortex briefly.

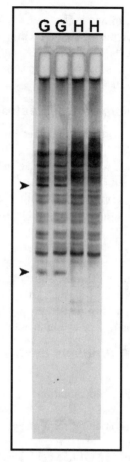

Fig. 2. Differential display of hydra genes from head and gastric cells. Lanes marked G contain PCR products derived from mRNA of gastric tissue. Lanes marked H contain PCR products derived from head specific cells. Arrowheads identify two genes specifically expressed in gastric tissue and absent in head specific cells.

4. Centrifuge at high speed in a microfuge for 30 s and boil for 2 min to support elution.
5. Vortex and centrifuge again.
6. Prepare four sterile PCR tubes containing 0 µL, 2 µL, 2.7 µL, and 3 µL of water.
7. Add eluation mix to the tubes to reach a final volume of 3 µL (in the last 3 µL tube; just dip the used pipet tip into the water).
8. From the elution mixtures, use the total of 3 µL of each dilution as template for a 10 µL PCR reaction. Mix 3 µL eluted PCR product, 1 µL lox PCR buffer, 1 µL random Homer primer (5 μM), 1 µL tailing primer (2 µM), 0.2 µL dNTP (10 mM), 0.1 µL Taq (0.5 U), 3.7 µL H_2O.

9. Cover reaction with 30 µL mineral oil.
10. PCR conditions: initial denaturing step 5 min at 94°C, 43 cycles of PCR with cycle times of 30 s at 94°C, 60 s at 42°C, and 30 s at 72°C.

3.6. Confirmation of Reamplification and High-Yield Second Reamplification (see Note 8)

1. For control of successful reamplification of the desired fragment, transfer 6 µL of the PCR reamplification mixture (not all!) into a new tube.
2. Freeze the remaining 4 µL of reamplification products at −20 °C.
3. Add 4 µL sequencing stop solution to the 6 µL aliquot.
4. Follow **steps 3.4** and **3.5** for loading reamplified fragments next to original PCR products (stored in −20°C).
5. If the PCR fragment of interest was successfully amplified, use an aliquot of the frozen PCR product (*see* **Step 2**) for second reamplification and further characterization (e.g., cloning and sequencing) (*see* **Note 9**).
6. To obtain large amounts of PCR product, e.g., for cloning, it is useful to perform a second reamplification step. Use the same concentrations as for first reamplification in a total volume of 100 µL.
7. As template use 2 µL of first reamplification product.
8. Purify PCR product by preparative agarose gel electrophoresis.

4. Notes

1. Critical for success in REN display experiments is to avoid both ribonuclease and nucleic acid contamination. Therefore, adopt semi-sterile techniques including wearing gloves and using water treated with 0.1% diethylpyrocarbonate.
2. Work quickly and keep all reagents on ice.
3. Concentration of cDNA-mixture is also essential for successful PCR.
4. Use dilutions of at least 1:25 to avoid interference with dNTPs and primers from the first strand cDNA synthesis mixture.
5. For silver-staining of cDNA use clean glassware for all incubations and ice-cold developer.
6. Note: 2% sodium thiosulfate stock solution is only stable for about one week at 4°C.
7. For elution of candidate bands the size of the gel slice used for elution should be as small as possible since it may contain toxic substances that inhibit the reamplification procedure.
8. After reamplification and cloning of the desired PCR fragment, screen multiple clones.
9. Any band recovered from the gel can contain more than one cDNA fragment. Thus, in order to detect the clone encoding the differentially expressed transcript, multiple cloned fragments have to be screened and analyzed by Northern blotting (*see* Chapter 15).

References

1. Hedrick, S. M., Cohen, D. I., Nielsen E. A., and Davis, M. M. (1984) Isolation of cDNA clones encoding T-cell specific membrane-associated proteins. *Nature* **308**, 149–153.

2. Travis, G. H. and Sutcliffe, J. G. (1988) Phenol emulsion-enhanced DNA-driven subtractive cDNA cloning: isolation of low-abundance monkey cortex-specific mRNAs. *Proc. Natl. Acad. Sci. USA* **85**, 1696–1700.

3. Dworkin, M. B. and Dawid, I. B. (1980) Construction of a cloned library of expressed embryonic gene sequences from *Xenopus laevis. Dev. Biol.* **76**, 435–448.

4. Lambolez, B., Audinat, E., Bochet, P., Crepel, F., and Rossier, J. (1992) AMPA receptor subunits expressed by single Purkinje cells. *Neuron* **9**, 247–258.

5. Liang, P. and Pardee, A. B. (1992) Differential display of eukaryotic messenger RNA by means of the polymerase chain reaction. *Science* **257**, 967–971.

6. McClelland, M., Mathieu-Daulde, F, and Welsh, J. (1995) RNA fingerprinting and differential display using arbitrarily primed PCR. *Trends Genet* **11**, 242–246.

7. Zimmermann, J. W. and Schultz, R. M. (1994) Analysis of gene expression in the preimplantation mouse embryo: use of mRNA differential display. *Proc. Natl. Acad. Sci. USA* **91**, 5456–5460.

8. Li, F., Barnathan, E. S., and Karikó, K. (1994) Rapid method for screening and cloning cDNAs generated in differential mRNA display: application of Northern blot for affinity capturing of cDNAs. *Nucl. Acids Res.* **22**, 1764–1765.

9. Watson, M. A. and Fleming, T. P. (1994) Isolation of differentially expressed sequence tags from human breast cancer. *Cancer Res.* **54**, 4598–4602.

10. Wong, K. K. and McClelland, M. (1994) Stress-inducible gene of Salmonella typhimurium identified by arbitrarily primed PCR of RNA. *Proc. Natl. Acad. Sci. USA* **91**, 639–643.

11. Liang, P., Averboukh L., Keyomarsi K., Sager R., and Pardee A. B. (1992) Differential display and cloning of messenger RNAs from human breast cancer versus mammary epithelial cells. *Cancer Res.* **52**, 6966–6968.

12. Bauer, D., Muller, H., Reich, J., Riedel, H., Ahrenkiel, V., Warthoe, P., and Strauss, M. (1993) Identification of differentially expressed mRNA species by an improved display technique (DDRT-PCR). *Nucl. Acid Res.* **21**, 4272–4280.

13. Lohmann, J., Schickle, H. P., and Bosch, T. C. G. (1995) REN, a rapid and efficient method for nonradioactive differential display and isolation of rnRNA. *BioTechniques* **18(2)**, 200–202.

14. Bassam, B. J., Caetano-Annolés, G., and Gresshoff, P. M. (1991) Fast and sensitive silver staining of DNA in polyacryamide gels. *Anal. Biochem.* **196**, 80–83.

24

Characterization of RNA Using Continuous RT-PCR Coupled with ELOSA

François Mallet, Guy Oriol, and Bernard Mandrand

1. Introduction

Characterization of RNA using Northern blotting or RNase protection assay is achieved by a two step procedure including a separation step and a detection step. Specificity and sensitivity are conferred by both the separation process and hybridization with radiolabeled probes. In both techniques, the observed results are validated after normalization of RNA performed by using a probe generally targeting a housekeeping gene. The greatest sensitivity was reached with the recently introduced reverse transcription-polymerase chain reaction (RT-PCR). However, because of the sensitivity of RT-PCR, false positives could present a problem. The selection of primer sets overlapping splice junctions could prevent amplification of potentially contaminating DNA in the RNA preparation. Using an RT-PCR process that does not require the reopening of the test tube could reduce the risk of introducing contaminating cloned cDNA or RNA during handling. False negatives could also present a problem. Quality of the RNA preparation (and quantity) could be checked by amplification of a housekeeping gene that informs one of the presence of potentially contaminating inhibitors of the enzymatic reaction. The specificity in RT-PCR assays is generally achieved using a detection analysis by Southern blot, dot blot, or restriction mapping. These techniques are generally used in specialized laboratories and could be replaced by microtiter plate or tube format assays well-adapted to clinical applications *(1)*. On such a format assay, PCR products must be captured on a solid phase and further detected by direct or indirect nonisotopic labeling (reviewed in **ref. 2**).

From: *Methods in Molecular Biology, Vol. 86: RNA Isolation and Characterization Protocols*
Edited by: R. Rapley and D. L. Manning © Humana Press Inc., Totowa, NJ

To address these problems, we have designed a procedure consisting of a continuous RT-PCR step coupled with a rapid and easy method for the nonradioactive detection on microtiter plates of RT-PCR-amplified nucleic acid sequences, which we call enzyme-linked oligosorbent assay (ELOSA). We developed a continuous RT-PCR protocol based on the use of Avian myelobastosis virus (AMV)-RT and *Taq* DNA polymerase, wherein all reaction components were added to a single tube prior to positioning in the thermocycler *(3)*. AMV-RT was chosen for reverse transcription because it has a longer half life and greater processivity than the MMuLV-RT, it can read through stable secondary mRNA structures at relatively high temperature *(4)*, and it can function under the same buffer conditions as *Taq* DNA polymerase *(5)*. Our cycling conditions included the linkage of the primary RT-PCR steps in rapid succession, which we considered an advantage since, as far as thermodynamics are concerned, achieving a rapid equilibrium between primer and template sequences substantially enhances the efficiency of hybridization *(6)*. Normalization of mRNA samples was performed by amplifying the aldolase-spliced mRNA. The continuous procedure was found to be specific in identifying splice sites of HIV-1 *(3)*, was sensitive in detecting at least 10 HIV-1 copies *(3)* or the hepatitis C RNA virus in serum *(7)*, and was reliable with an error rate of less than 0.2% *(3)*. The detection method we developed is based on the sandwich hybridization, in solution and at 37°C, between single-strand DNA template and short synthetic oligonucleotides (18–27 mer) as capture and detection probes *(8)*. The fixation of the capture oligonucleotides was done by a passive adsorption on the surface of the microtiter wells, using phosphate buffer saline containing a high concentration of salts. The detection probe is a horseradish peroxidase-labeled oligonucleotide. ELOSA was shown to be convenient as a routine assay and was applied to HIV-1 detection *(8,9)* and quantization *(10)*, Human leukocyte antigen (HLA) typing *(11)*, and automated *Mycobacterium tuberculosis* detection *(12)*.

2. Materials

2.1. Amplification

1. 1 µg/5 µL total RNA in water.
2. Diethylpyrocarbonate (DEPC)-treated water. Prepare by adding 1 mL of DEPC to 1L of H_2O. Mix the solution thoroughly and allow to set 30 min to overnight followed by autoclaving (*see* **Note 1**).
3. 23.4 U/µL RNAguard (Pharmacia Biotech, Pharmacia Biotech, Uppsala, Sweden) (*see* **Note 2**).
4. 10X amplification buffer: 100 m*M* Tris-HCl pH 8.3, 500 m*M* KCl, 0.1% gelatin; (*see* **Note 3**).
5. 100 m*M* $MgCl_2$ in sterile water (Merck, Darmstadt, Germay) (*see* **Note 3**).

6. 100 mM 4dNTP mix: 25 mM each dNTP, Pharmacia LKB Biotechnology; combination of equal volumes of commercially available 100 mM stock solutions of dATP, dCTP, dGTP, dTTP.
7. 31 pmol/µL 5' primer in H_2O (*see* **Note 4**).
8. 31 pmol/µL 3' primer in H_2O (*see* **Note 4**).
9. *Taq* DNA polymerase 5 U/µL (Perkin Elmer, Branchburg).
10. AMV reverse transcriptase 25 U/µL (Boehringer Mannheim, Mannheim, Germany) (*see* **Note 5**).
11. Mineral oil (Sigma, St. Louis, MO).

2.2. Detection

1. Microtiter plates (Maxisorb, NUNC) (*see* **Note 6**).
2. PBS 10X: 1.37 M NaCl, 27 mM KCl, 43 mM Na_2HPO_4, 14 mM $KH_2PO_4.$
3. Adsorption buffer: PBS 3X.
4. Washing buffer: PBS 1X, 0.05% Tween-20 (w/v).
5. Hybridization buffer: 0.1 M sodium phosphate, pH 7.0, 0.5 M NaCl, 0.65% Tween-20 (w/v), 0.14 mg/mL salmon sperm DNA (Boehringer Mannheim), 2% PEG 4000 (*see* **Note 7**).
6. Substrate: OPD (*O*-phenylenediamine, Sigma) 2 mg/mL in 0.05 M citric acid, 0.1 M Na_2HPO_4, pH 4.9, containing H_2O_2 (0.03 vol) (*see* **Note 8**).
7. Denaturation buffer: 2 M NaOH.
8. Neutralization buffer: 2 M Acetic acid.
9. 1N H_2SO_4 (Merck).
10. Capture probe (150 nM in adsorption buffer) (*see* **Note 9**).
11. Detection probe (15 nM in hybridization buffer) (*see* **Note 9**).

3. Methods
3.1. General Considerations

The following guidelines are recommended in order to eliminate the possibility of contamination. Several dedicated areas required; fresh disposable gloves should be used in each area. A first area is dedicated to the sample treatment; no plasmid preparation or RT-PCR reaction products should enter this area. A second area is dedicated to the preparation of the reaction mixture; no plasmid preparation, no RT-PCR reaction products and no template should enter this area. The use of a laminar flow hood equipped with UV light is effective as a clean area. This area should have dedicated pipets and pipet tips. Sterile microcentrifuge tubes and pipet tips should be used for all reactions. These items should be sterilized by steam autoclaving with the tubes sealed. Sterile, cotton-plugged pipet tips are recommended to prevent carryover between samples. Pipet tips should not be used more than once. Tubes are sealed between each addition of reagents. Controls with no template should be set up entirely in the clean area and not reopened until ready for analysis. A third area is specifically

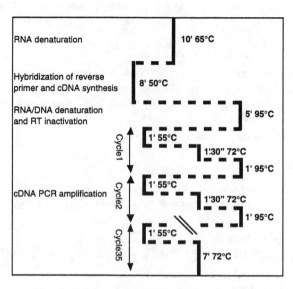

Fig. 1. Scheme of the RT-PCR procedure.

dedicated to the addition of the template to the reaction mixture; no plasmid preparation or RT-PCR reaction products should enter this area. Care should be taken not to cross-contaminate gloves during template addition. Tubes are handled one by one and sealed immediately after addition of template. The last dedicated area is the regular laboratory where the detection procedure occurs.

3.2. RNA Preparation

Total RNA or poly(A)⁺ RNA are prepared using rapid or classical guanidinium isothiocyanate-based protocols.

3.3. Continuous RT-PCR Reaction

Reaction conditions have been optimized for the Perkin Elmer 480A apparatus. The principle of RT-PCR procedure is summarized in **Fig. 1**. The reaction master mix is prepared in its dedicated area. Master reagent mixtures should be prepared in excess to allow for pipeting losses. As a rule of thumb, at least 10% extra volume is recommended. When adding reagents, pipet up and down several times to mix. For 10 standard RT-PCR tests, prepare a master mix for 11 reactions by adding the following to a sterile 1.5 mL microcentrifuge tube:

1. DEPC-treated water: 864.6 µL.
2. RNAguard: 11 µL.

3. 10X amplification buffer: 110 µL.
4. MgCl$_2$ (*see* **Note 10**): 16.5 µL.
5. 4dNTP mix: 11 µL.
6. 5' primer: 11 µL (*see* **Note 11**).
7. 3' primer: 11 µL (*see* **Note 11**).
8. *Taq* DNA polymerase (*see* **Note 12**): 5.5 µL.

The following **steps (9–11, 14)** are performed on ice.

9. AMV-RT (*see* **Notes 12, 13**): 4.4 µL.
10. Gently mix by hand and spin briefly in a microcentrifuge to ensure all contents are in the bottom of the tube (*see* **Note 14**).
11. Distribute 95 µL of the master mix to the bottom of each reaction tube.
12. Cover the RT-PCR master mix with two drops of mineral oil.
13. Move to the area dedicated to template addition.
14. Add 5 µL of total RNA (1 µg) in water (*see* **Note 15**) to the RT-PCR master mix. This may be accomplished by placing the pipet tip through the oil, and pipeting the RNA solution into the lower aqueous phase. Pipet carefully up and down several times.

    The final concentrations of components after RNA is added are 1X amplification buffer, 0.234 U/µL RNAguard, 1.5 mM MgCl$_2$, 250 µM each dNTP, 310 nM each primer, 0.025 U/µL of *Taq* DNA polymerase and 0.1 U/µL of AMV-RT.
15. Gently mix by tapping the tube, and spin briefly in microcentrifuge to obtain a clean oil/aqueous interface and to remove air bubbles.
16. Amplify by RT-PCR using the following cycle profile:
    65°C, 10 min (RNA denaturation) (*see* **Note 16**)
    50°C, 8 min (reverse transcription) (*see* **Note 17**)
    95°C, 4 min (DNA/RNA denaturation and RT inactivation)
    95°C, 1 min –55°C, 1 min –72°C, 1.5 min (× 35 cycles)
    72°C, 7 min (final extension)
    8°C (hold).

3.5. Detection of RT-PCR Products by ELOSA

After amplification, move to the area dedicated to detection. The principle of ELOSA detection procedure is summarized in **Fig. 2**. The use of a multichannel pipet is highly recommended.

1. Drop 100 µL of PBS 3X hybridization buffer containing 150 nM capture probe in each well of the microtiter plate.
2. Incubate 2 h at 37°C or overnight at room temperature.
3. Wash the plate three times with PBS 1X, 0.05% Tween-20 (w/v) (*see* **Note 18**).
4. Dilute 25 µL out of 100 µL of the amplified double stranded DNA in 65 µL of hybridization buffer (*see* **Note 19**).
5. Add 10 µL of denaturation buffer (NaOH 0.2 M final concentration) (*see* **Note 19**).
6. Incubate at room temperature for 5 min.

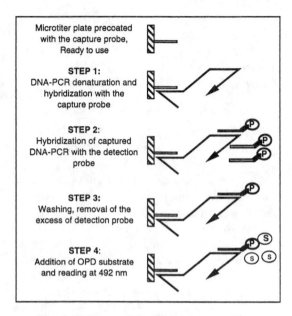

Fig. 2. Scheme of the ELOSA procedure.

7. Add 10 µL of neutralization buffer (acetic acid 0.2 *M* final concentration) (*see* **Note 19**).
8. Adjust the volume to 125 µL using the hybridization buffer (*see* **Note 20**).
9. Drop 50 µL of this solution in a well of a microtiter plate previously coated with a specific capture probe. Test each sample in two separate wells.
10. Immediately thereafter, add 50 µL of the detection probe (15 n*M* in hybridization buffer) to each well.
11. Incubate for 1 h at 37°C.
12. Wash the microtiter plate three times with washing buffer.
13. Add 100 µL of OPD substrate to each well.
14. After 30 min, stop the reaction by the addition of 100 µL of 1 N H_2SO_4.
15. Read the signal at A492 nm in an automatic microtiter plate reader.

4. Notes

1. Classical precautions concerning RNA handling should be taken, particularly the use of DEPC-treated solutions. Although residual DEPC has been described to inhibit *Taq* DNA polymerase, no effect was observed in RT-PCR using master mix prepared with or without DEPC-treated reagents.
2. RNAguard was not systematically used, and no differences were observed.
3. Tris-HCl, KCl, and gelatin (Merck) are weighed, dissolved in DEPC-treated water with slight heating to solubilized gelatin, and filtered through a 0.22-µm filter (Millipore). $MgCl_2$ stock solution is either autoclaved or filtered using a

0.22-μm filter. The use of commercially available, 10X premade amplification buffers is recommended; generally they are purchased with 15 mM MgCl$_2$ included. MgCl$_2$ premade solutions are also commercially available.

4. Oligonucleotides are synthesized on an Applied Biosystems 394 synthesizer by the phosphoramidite method, and generally do not require further purification. Primers are chosen empirically by the research worker. Duplex and hairpin formation, false priming sites, and quality of primers are checked using commercial programs (OLIGO4.03, National Bioscience, Plymouth, MN). Pay attention to choose primers with a difference between T$_m$ generally lower than 3°C. Overlapping exon primers are designed with the 3' end that covers the inner exon representing 35% of the primer T$_m$. It is possible to substitute inosine for point mutations when required. For RNA normalization using the aldolase system, prepare 6 pmol/μL primer stock solutions (*see* **Note 11**).

5. The source of AMV-RT appears not to be important, since the Boehringer Mannheim, Seikagaku, and Gibco-BRL enzymes all worked well in the procedure. MMuLV-RT, RNase-H plus or minus, did not show the same activity in the continuous RT-PCR assay.

6. The use of irradiated plates is highly recommended so as to increase adsorption efficiency.

7. The hybridization buffer, passed through a 0.22-μm filter and sterilized by steam autoclaving, is stable for at least 6 mo when stored at 4°C.

8. The substrate solution is prepared extemporaneously.

9. The oligonucleotides used for the detection and capture are synthesized with an amine arm at the 5' end. The addition was performed on a synthesizer with the aminolink II reagent (Applied Biosystems). The horseradish peroxidase labeling of detection oligonucleotides is performed as described by Urdea et al. *(13)*. Preparation of the oligonucleotide-enzyme conjugate is summarized as follows: to conjugate oligonucleotides to horseradish peroxydase (HRP), 400 μg of vacuum-dried 5-aminoalkyl-oligodeoxyribonucleotide were mixed first with 25 μL of 0.1M sodium borate buffer (pH 9.3) and then with 1 mL of 30 mg/mL 1,4-phenylene diisothiocyanate (DITC, Sigma) prepared in dimethylformamide. The mixture was incubated in the dark for 2 h at room temperature and transferred in a glass tube (corex 30-mL), then 5 mL of H$_2$O saturated-1-butanol were added and mixed, then 3 mL of H$_2$O were added. After three extractions with equal volumes of H$_2$O saturated-1-butanol, the aqueous phase was vacuum-dried and activated oligonucleotides resolubilized in 400 μL of 0.1 M sodium-borate buffer (pH 9.3) containing 9.2 mg of HRP (Boehringer). The mixture was incubated overnight at room temperature and then can be stored several days at 4°C. Oligonucleotide-HRP conjugates were HPLC purified by ion exchange chromatography, using WATERS Protein-Pak DEAE 8HR column and 10 to 42% buffer B gradient (buffer A: Tris-HCl 20 mM; pH 7.6; buffer B: Tris-HCl 20 mM; NaCl 2 M; pH 7.6). The oligonucleotide HRP-containing fraction is immediately dialyzed against H$_2$O overnight at 4°C, then vacuum-dried and mixed with 500 μL Tris 50 mM NH$_4$Cl 100 mM, pH 7.5. The purity of the

detection probe is greater than 95% and the ratio (oligonucleotide mole/L)/(HRP mole/L) is between 0.9 and 1.3. Detection probes, in Tris 50 mM NH$_4$Cl 100 mM, pH 7.5, 1 mg/mL salmon sperm DNA, 50% glycerol, are stable several mo when stored at –20°C. The HRP-probe concentration must be greater than 150 nM. Probes are chosen and checked like primers (*see* **Note 4**). It is possible to substitute inosine for point mutations when required. Such labeled probes are commercially available.

10. As Mg^{2+} concentration is a critical factor in the stringency of primer annealing, it is the first factor to vary when beginning a study to optimize the specificity and sensitivity of continuous RT-PCR. Final Mg^{2+} concentration can vary from 1.5 to 6 mM without altering the activities of both the AMV-RT and *Taq* DNA pol enzymes *(3)*.

11. Normalization of 1 μg RNA is achieved by amplifying aldolase mRNA using standard RT-PCR protocol except 60 nM each primer is used so as not to reach the plateau (primers are described in **Table 1**).

12. The AMV-RT: *Taq* DNA polymerase ratio affects the sensitivity of the continuous RT-PCR method. It is the second factor to vary when beginning a study to optimize sensitivity. AMV-RT: *Taq* DNA polymerase ratio can range from 5:2.5 to 20:2.5, although a significant decrease in signal has been observed at 20:2.5 when using less than 500 copies of template *(3)*.

13. One advantage of this two enzyme system is that you can omit the RT but perform exactly the same protocol, including cycling. This is particularly useful to control the absence of contaminating DNA when exon-spanning primers are not available.

14. Quantitative competitive RT-PCR is based on the coamplification of different amounts of competitive mutated RNA template along with the RNA target *(14)*. The mimic template differs from the wild-type target by only some mutations. For QC-RT-PCR purpose, the master mix is split into four parts; different amounts of competitive RNA template are added outside the master mix dedicated area. The standard protocol is applied, except 45 PCR cycles are performed in order to reach the plateau.

15. 1 μg RNA is used in standard normalized procedure (*see* **Table 2** and **Fig. 3**). However, either a crude RNA preparation or poly(A)$^+$ RNA can be used. The RT-PCR procedure achieves the same level of amplification from 50 copies of target, whether diluted in 10 ng or 1 μg of nonspecific RNA, in buffer conditions that included 1.5 mM Mg^{2+}, and at an AMV-RT to *Taq* DNA polymerase ratio of 10:2.5.

16. Depending on the required sensitivity and the complexity of the target, the temperature used during the RNA denaturation step can vary from 60°C to 75°C.

17. Depending on the required specificity and the complexity of the target, the reverse transcription step can vary from 5 min at 40°C to 15 min at 65°C. A short step such as 5 min at 37°C can be introduced between the standard denaturation step (10 min at 65°C) and the reverse transcription step (8 min at 50°C), so as to allow non-stringent hybridization of reverse primer, or when using degenerate primers. The hybridization temperature in the following PCR should be subsequently modified.

Table 1
Primers and Capture and Detection Probes[a]

Target gene	Oligonucleotide function	DNA sequence (5' to 3')	Name	Location
Aldolase	PCR	CCCCTTCCGAGGCTAAATCG	U1179	2721-2740
	PCR (reverse)	CTGGTAGTAGCAAGTTC-CTGGCAC	L1178	4177-4161 to 2866-2860
	Capture	CCTTGAATCCACTCGCCAGCC	C4031	2796-2816
	Detection	GCAGAAGGGGTCCTGGTGACG	D4034	2760-2780
HIV-1 MS-mRNA	PCR	TCTAICAAAGCA-ACCCIC	U1082	5580-5591 to 7925-7930
	PCR (reverse)	CCTATCTGCCCCTCAGCTAC	L972	8254-8234
	Capture	GACCCGACAGGCCCGAAGGAATC	C4033	7947-7969
	Detection	CTACCACCGCTTGAGAGACTTACT	D4035	8071-8094

[a] The *Aldolase* positions are with reference to the HSALDOA sequence (GenBank). HIV-1 nomenclature is according to HIVHXB2R (15).

169

Table 2
RT-PCR-ELOSA Normalization of RNA[a]

Sample	Mean OD	CV (%)
1	1768	6.6
2	1631	7.5
3	1664	5.7
4	1227	4.2
5	1166	3.0
6	1430	5.3
All	1481	4.1

[a]Six independant RT-PCR amplifications of 1 μg total RNA were performed for six samples and aldolase amplification products were detected by ELOSA. ODs at 492 nm are expressed as OD × 1000. Means and CVs are indicated. The acceptance criterion for normalization of 1 μg total RNA was fixed at an OD between 700 and 2200, defined by mean ± 2 SD of the 36 RT-PCR-ELOSA (All).

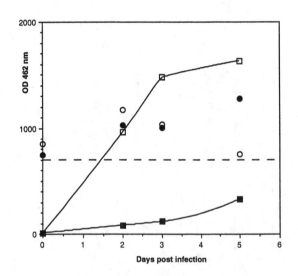

Fig. 3. Kinetic of HIV-1 infection in vitro. A rapid-high virus (RH-virus) and a slow-low virus (SL-virus) were used. HIV-1 viral stocks with an RT activity of 10^6 cpm were mixed with 2×10^8 PBMC during 2 h, then the infected cells were washed and grown in fresh media. Each day 3×10^7 cells were collected and total RNA was extracted. RNAs (1 μg) were normalized by amplifying the aldolase RNA (○ and ●) and the multiply-spliced HIV-1 mRNA were detected (□ and ■), for the RH-virus (open symbols) and SL-virus (black symbols) infected cells, respectively. ELOSA was performed using 1/75th of the RT-PCR solution (*see* **Note 20**). ODs at 492 nm are expressed as OD × 1000. Acceptance criterion for RNA quality is shown by a dot line (*see* **Table 2**). Primers and probes are described in **Table 1**.

18. Oligonucleotide-coated plates, dried in a vacuum oven and sealed by a plastic sheet, are stable for several months at 4°C.
19. This step is a crucial one in ELOSA. Do not try to lower the input of PCR solution under 20 µL and always complete to 90 µL with hybridization buffer. Do not modify the volumes in **steps 5** and **7**. Modifications would decrease denaturation efficiency and therefore hybridization efficiency.
20. The volume of added hybridization buffer could vary depending of the goal of the research worker. For sensitive detection purpose, a yes/no response is expected, then OD signals can reach the saturation of the microtiter reader. Adjusting the volume to 125 µL is adequate for detection beneath the 10 copies level (the input of 50 µL in a well is equivalent to 10 µL of the initial RT-PCR solution, that is 1/10th). For normalization of RNA using the aldolase system, adjust the volume to 935 µL using the hybridization buffer (input equivalent to 1/75th of the RT-PCR solution) (*see* **Fig. 3**). Correct normalization is achieved when the measured OD range between a minima and a maxima defined as acceptance criteria (*see* **Table 2** and **Fig. 3**). Acceptance criteria should be defined according to the source of the sample, the nature of the sample, and sample treatment *(10)*. For quantization of mRNA using QC-RT-PCR, a higher dilution is required (input equivalent to 1/200th of the RT-PCR solution) *(14)*; wild type and mimic templates are discriminated by ELOSA at the capture probe level *(10,14)*.

Acknowledgments

The authors wish to thank Christelle Brun and Nathalie Ferraton for oligonucleotides synthesis and enzyme coupling, C. Guillon for gift of HIV-1 RNA, and P. Cros and A. Laayoun for helpful discussions.

References

1. Whetsell, A. J., Drew, J. B., Milman, G., Hoff, R., Dragon, E. A., Adler, K., Hui, J., Otto, P. Gupta, P., Farzadegan, H., and Wolinsky, S. M. (1992) Comparison of three nonradioisotopic polymerase chain reaction-based methods for detection of human immunodeficiency virus type 1. *J. Clin. Microbiol.* **30**, (4) 845–853.
2. Manak, M. M. (1993) Hybridization formats and detection procedures. In Keller, G. H. and Manak, M. M., (eds.) *DNA Probes.* Stockton, New York. pp. 199–249.
3. Mallet, F., Oriol, G., Mary, C., Verrier, B., and Mandrand, B. (1995) Continuous RT-PCR using AMV-RT and Taq DNA polymerase: characterization and comparison to uncoupled procedures. *BioTechniques* **18**, 678–687.
4. Shimomaye, E. and Salvato, M. (1991) Use of avian myeloblastosis virus reverse transcriptase at high temperature for sequence analysis of highly structured RNA. *Gene Anal. Tech.* **6**, 25–28.
5. Wang R. F., Cao, W. W., and Johnson, M. G. (1992) A simplified, single buffer system for RNA-PCR. *BioTechniques* **12**, 702–704.
6. Hearst, E. H. (1988) A photochemical investigation of the dynamics of oligonucleotide hybridization. *Ann. Rev. Phys. Chem.* **39**, 291–315.

7. François M., Dubois, F., Brand, D., Bacq, Y., Guerois, C., Mouchet, C., Tichet, J., Goudeau, A., and Barin, F. (1993) Prevalence and significance of hepatitis C (HCV) viremia in HCV antibody-positive subjects from various populations. *J. Clin. Microbiol.* **31**, 1189–1193.

8. Mallet, F., Hebrard, C., Brand, D., Chapuis, E., Cros, P., Allibert, P., Besnier, J. M., Barin, F., and Mandrand, B. (1993) Enzyme-linked oligosorbent assay for detection of polymerase chain reaction-amplified human immunodeficiency virus type 1. *J. Clin. Microbiol.* **31**, 1444–1449.

9. Brossard, Y., Aubin, J. T., Mandelbrot, L., Bignozzi, C., Brand, D., Chaput, A., Roume, J., Mulliez, N., Mallet, F., Agut, H., Barin, F., Brechot, C., Goudeau, A., Huraux, J. M., Barrat, J., Blot, P., Chavinie, J., Ciraru-Vigneron, N., Engelman, P., Hervé, F., Papiernik, E., and Henrion, R. (1995) Frequency of early *in utero* HIV-1 infection: a blind DNA polymerase chain reaction study on 100 fetal thymuses. *AIDS* **9** (4), 359–366.

10. Mallet, F., Hebrard, C., Livrozet, J. M., Lees, O., Tron, F., Touraine, J. L., and Mandrand, B. (1995) Quantitation of human immunodeficiency virus type 1 DNA by two PCR procedures coupled with enzyme-linked oligosorbent assay. *J. Clin. Microbiol.* **33**, 3201–3208.

11. Cros, P., Allibert, P., Mandrand, B., Tiercy, J. M., and Mach, B. (1992) Oligonucleotide genotyping of HLA polymorphism on microtitre plates. *Lancet* **340**, 870–873.

12. Mabilat, C., Desvarenne, S., Panteix, G., Machabert, N., Bernillon, M. H., Guardiola, G., and Cros, P. (1994) Routine identification of *Mycobacterium tuberculosis* complex isolates by automated hybridization. *J. Clin. Microbiol.* **32**, 2702–2705.

13. Urdea, M. S., Warner, B. D., Running, J. A., Stempien, M., Clyne, J., and Horn, T. (1988) A comparison of nonradioisotopic hybridization assay methods using fluorescent, chemiluminescent and enzyme labeled synthetic oligodeoxyribonucleotide probes. *Nucleic Acids Res.* **16** 4937–4956.

14. Mallet, F. (1996) Continuous RT-PCR using AMV-RT and *Taq* DNA polymerase. In: Burke, J. (ed) *PCR: Essential Techniques.* BIOS Scientific, Pp. 82–85, J. Wiley, Chichester.

15. Myers G., Korber, B., Berzofsky, J. A., Smith, R. F., and Pavlakis, G. N. (1991) Human retroviruses and AIDS: a compilation and analysis of nucleic acid and amino acid sequences. IA3-IA69. Los Alamos National Laboratory, Los Alamos, NM.

25

Gene Expression Analysis by CD-RT-PCR

Eric de Kant

1. Introduction
1.1. Gene Expression Analysis by PCR

With the introduction of polymerase chain reaction (PCR), analysis on even minute amounts of DNA and RNA from biological samples became practicable. Conventional methods of mRNA analysis are often not sensitive enough for detection of low-abundance transcripts or broad examination of gene expression in small amounts of RNA. Quantitative mRNA characterization by reverse transcription (RT) of RNA and subsequent PCR (RT-PCR) is, however, compared to qualitative RT-PCR detection of RNA, more complicated because of two features inherent in in vitro amplification. First, during the exponential phase, minute differences in a number of variables can greatly influence reaction rates, with substantial effect on the yield of PCR products. Second, as a consequence of the consumption of reaction components and generation of inhibitors, the amplification enters a plateau phase. At this point, the reaction rate declines to an unknown level. Another source of errors in quantitative RT-PCR analysis lies in the determination of the amount of RNA to be analyzed for each sample. In small samples, the total amount of RNA may even be beyond the limit of detection. The sample loading problem can be solved by presenting the level of expression of the gene of interest in reference to a constitutively expressed gene. In PCR, this can be done by the simultaneous amplification of two different genes in one reaction vessel, which is called differential PCR *(1)*. However, in many cases a quantitative PCR assay is desired that is internally controlled both for errors in comparison between samples and for the efficiency of the amplification reaction. To this end, a tech-

From: *Methods in Molecular Biology, Vol. 86: RNA Isolation and Characterization Protocols*
Edited by: R. Rapley and D. L. Manning © Humana Press Inc., Totowa, NJ

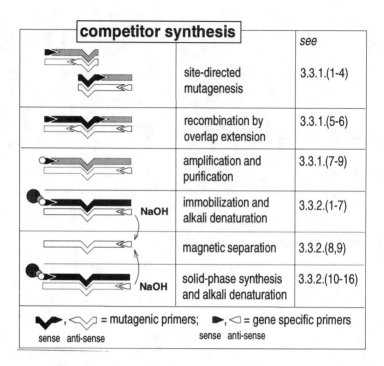

Fig. 1. A schematic outline of antisense competitor DNA synthesis.

nique was devised that combines competitive PCR and differential RT-PCR by co-amplification of two genes and their corresponding competitive templates *(2)*. This chapter describes the working procedures for that complete assay, called competitive and differential RT-PCR (CD-RT-PCR) and concomitant techniques.

1.2. Synthesis of Mutant Antisense Competitive DNA Fragments

CD-RT-PCR requires using site-specific mutagenized single-stranded DNA fragments as competitors *(2)*. These competitive DNA fragments can be easily generated without subcloning (**Fig. 1**). For that purpose, two successive rounds of PCR are performed for both genes. In the first, mutagenic primers are used in two separate PCRs to create or remove unique restriction sites. The resulting two mutant fragments, which are homologous around the mutated sequence, are recombined by overlap extension in the second round of PCR *(3)*. The biotinylated sense strands can be physically separated from the required antisense strands by alkali denaturation following immobilization of the biotinylated PCR fragments that are directly bound to streptavidin-coupled polystyrene magnetic beads.

1.3. Outline of the Complete CD-RT-PCR Assay

A schematic outline of the concept of CD-RT-PCR is shown in **Fig. 2.** CD-RT-PCR comprises random-primed reverse transcription followed by co-amplification of first-strand cDNAs from both the gene of interest and the reference gene, and their matching artificially mutated antisense competitive DNAs. The competitive templates are mutagenized in a way that either the competitor DNA or the cDNA sequence contains a unique restriction site exactly in the center of the segment that is amplified during CD-RT-PCR. The high homology between wild-type and mutant sequences promotes perfect competition not only during PCR but also during an essential final re-annealing step. Note that only mutant antisense (not double-stranded) DNA acts as perfect competitor for first-strand cDNAs *(2)*. After CD-RT-PCR, restriction enzyme digestion enables visualization of digestible homoduplexes and nondigestible homo- and heteroduplexes. Separation of the digested DNA by gel electrophoresis or HPLC results in a pattern of four different fragment sizes. HPLC analysis provides the quantitative information concerning the ratios of digested and undigested amplification products that is required for calculation of the amounts of the two specific cDNAs prior to PCR and finally the normalized expression level of the gene of interest.

1.4. Summary

CD-RT-PCR combined with discrimination of amplified cDNAs and competitor DNAs via restriction enzyme digestion and HPLC analysis provides a sensitive assay for reproducible and accurate measurement of normalized gene expression without the need for extensive competitor titrations for every single sample.

2. Materials

1. Cells or tissue from which RNA will be extracted.
2. RNA purification kit: RNAzol (Biotecx Laboratories, Houston, TX).
3. SuperScript RNase H–Reverse Transcriptase (200 U/µL; Life Technologies, Gaithersburg, MD) with manufacturer-recommended buffer and DTT (0.1M).
4. rRNasin ribonuclease inhibitor (40 U/µL; Promega, Madison, WI).
5. Random Primer oligonucleotides (hexamers; Life Technologies).
6. A mixture of 2.5 mM of each deoxynucleotide (4dNTP mix: dATP, dGTP, dCTP, and dTTP).
7. 10X PCR buffer: 100 mM Tris-HCl, pH 8.3, 500 mM KCl, 15 mM MgCl$_2$, 0.01% (w/v) gelatin.
8. AmpliTaq DNA Polymerase (5 U/µL; Perkin-Elmer).
9. Two sets of gene-specific primers (10 µM): A nonmodified and a biotinylated primer complementary to the 5' end (sense primer) and a nonmodified primer complementary to the 3' end (antisense primer) of the sequence to be amplified

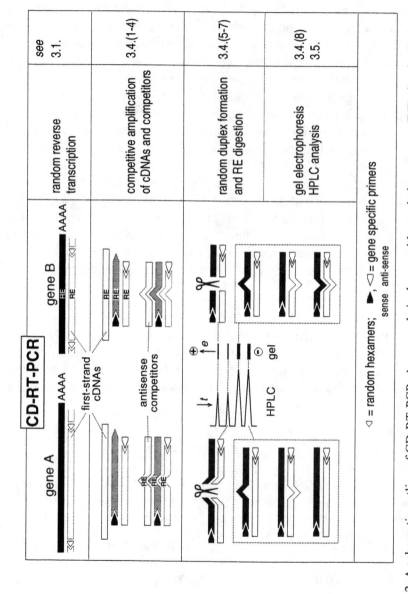

Fig. 2. A schematic outline of CD-RT-PCR. An example is shown with restriction enzyme (RE) sites in the competitor of gene A and in the wild-type cDNA sequence of gene B.

during CD-RT-PCR; mutagenic (sense and antisense) primers for preparation of a mutant fragment with a newly created or removed restriction enzyme recognition site. Mutagenic primers are complementary to the central portion of the CD-RT-PCR fragments (*see* **Note 1** and **Fig. 1**).

10. Light mineral oil.
11. An automated thermal cycler (e.g., the author uses a Perkin-Elmer machine).
12. GeneClean (Bio101, La Jolla, CA), or any other method for purifying DNA from agarose gels.
13. Streptavidin-coated magnetic beads: Dynabeads M-280 Streptavidin, and a Magnetic Particle Concentrator: Dynal MPC for Eppendorf tubes (Dynal, Oslo, Norway).
14. 2X Binding and Washing buffer (B&W): 10 mM Tris-HCl, pH 7.5, 1 mM EDTA, 2M NaCl.
15. TE buffer: 10 mM Tris-HCl, pH 8.0, 1 mM EDTA.
16. 0.1N NaOH for alkali denaturation, 1N HCl for neutralization, and 1M Tris-HCl (pH 8.0) for buffering.
17. Siliconized tubes.
18. Carrier DNA (10 ng/μL): Sonicated Salmon Sperm DNA (Stratagene, La Jolla, CA)
19. Antisense competitor DNA (*see* **Subheadings 3.3.** and **3.4.**).
20. Appropriate restriction enzyme for digestion of CD-RT-PCR products (*see* **Note 1**).
21. Potassium glutamate buffer (6X KGB): 600 mM potassium glutamate, 150 mM Tris-acetate (pH 7.5), 60 mM magnesium acetate, 300 μg/mL bovine serum albumin (Fraction V; Sigma), 3 mM ß-mercaptoethanol.
22. Materials for gel electrophoresis and staining of DNA *(4,5)*: Electrophoresis buffer; 6X nondenaturing gel-loading buffer and 95% formamide-gel-loading buffer; EDTA/SDS solution: 10 mM EDTA, 0.1% SDS; DNA mol w marker; DNA grade agarose (normal and low-melting-point agarose); acrylamide, *bis*-acrylamide, ammonium persulfate and TEMED (*N,N,N',N'*-Tetramethyl-ethylenediamine) for 10% nondenaturing polyacrylamide gels; ethidium bromide; Silver Stain Kit (Bio-Rad, Richmond, CA).
23. An automated HPLC system. In this chapter the chromatographic system is from Knauer and consists of a solvent delivery system, a UV detector (260 nm), and an autosampler equipped with a 20 μL loop. Data were collected using a Shimadzu integrator.
24. Analytical anion-exchange DEAE-NPR column (Perkin-Elmer), 35 mm × 4.6 mm I.D., packed with 2.5 μm particles.
25. HPLC buffer A: 1M NaCl, 25 mM Tris-HCl, pH 9.0, and buffer B: 25 mM Tris-HCl, pH 9.0.

3. Methods

3.1. RNA Purification and First-Strand cDNA Synthesis

All manipulation must be carried out in an RNase free environment.

1. Isolate RNA using the RNAzol method or any other established procedure (**ref. *6***, *see* Chapter 1–10) depending on the type of tissue. RNAs must be stored at –80°C.

2. Combine the following to perform a 20 μL reverse transcriptase (RT) reaction:
   RNA (0.2 to 2 μg; *see* **Note 2**)
   2 μL 10X PCR buffer (results in equal or slightly higher yields than 4 μL
      5X RT buffer)
   2 μL 0.1*M* DTT
   1 μL 0.1 m*M* random primers (*see* **Note 3**)
   5 μL 2.5 m*M* 4dNTP mix
   Ultra pure H_2O to 18 μL
3. Denature RNA at 94°C for 3 min in this mixture and cool on ice.
4. Add 1 μL RNasin and 1 μL reverse transcriptase (2 μL of a 1:1 mixture).
5. Incubate at room temperature for 10 min and at 42°C for 40 min.
6. Heat reaction to 94°C for 3 min and chill on ice for 1 min to inactivate the enzyme
   and to denature the nucleic acids.
7. Proceed directly to PCR (**Subheading 3.2.**) or store the cDNA at –20°C.

3.2. Basic Protocol for PCR Amplification

The conditions for PCR amplification of DNA depend on the combination
of template and primers that is used. A few variables may have to be optimized
to achieve a fair result with high enough sensitivity and specificity: $MgCl_2$,
template and primer concentrations, and cycle profile parameters. To allow
detection of contaminations, always prepare a blank PCR that contains all
reaction components but DNA.

1. Combine the following in a 500 μL tube to perform a 25 μL PCR:
   Template DNA: cDNA representing 10–100 ng RNA or amplification
      products (e.g., competitors) derived from a previous PCR (1 fg to 1000 pg),
   2.5 μL 10X PCR buffer,
   2 μL 2.5 m*M* 4 dNTP mix,
   0.1 to 5 μL 10 μ*M* oligonucleotide primers (0.01–0.5 μ*M* final;
      see **Subheading 3.4.**),
   H_2O to 24.9,
   0.1 μL *Taq* DNA polymerase (*see* **Note 4**).
2. Overlay reaction with mineral oil.
3. Carry out 25 to 35 PCR cycles in an automated thermal cycler under conditions
   that are best suited for each specific type of PCR. A typical cycle profile using
   PCR primers of about 20 bases in length with a GC content close to 50% yielding
   amplification products of <500 bp (*see* **Note 1**) may look like this:
      5 min, 94°C (initial denaturation),
      Followed by 25 to 35 cycles of:
      30 s, 94°C (denaturation),
      30 s, 50 to 60°C (annealing),
      30 s, 72°C (extension),
      Followed by:
      10 min, 72°C (final extension).

4. Remove PCR product by inserting a micropipette through the mineral oil layer and drawing up the sample.
5. Analyze 5 to 10 µL of the reaction mix by agarose gel electrophoresis to verify the size, purity and yield of PCR products and/or analyze by HPLC (*see* **Subheading 3.6.**).

3.3. Synthesis of Antisense Competitor DNA for CD-RT-PCR

3.3.1. Site-Directed Mutagenesis and Recombination of cDNA by PCR

The purpose of this PCR technique *(3)* is to mutagenize the center of a cDNA fragment, that will afterwards serve as competitive template in CD-RT-PCR, so that a unique restriction enzyme recognition site is either created or removed from that sequence segment. To avoid time-consuming DNA purification, the mutant products of a first round of PCR are excised from low-melting-point agarose and directly used for recombination in a second round of PCR. The protocol is schematically outlined in the first part of **Fig. 1** and applies to both the gene of interest and the reference gene. PCR is carried out basically as described in **Subheading 3.2.**

1. Perform a first round of two separate PCRs, each with a mutagenic and a normal primer, to create two partly overlapping mutated amplification products.
2. Separate mutant amplification products by gel electrophoresis on a preparative low-melting-point agarose gel.
3. Locate bands of interest using UV-illumination of the ethidium bromide stained gel (*see* **Note 5**).
4. Excise bands of interest with a razor blade and transfer to a microfuge tube.
5. Melt gel slices at 68°C (5–10 min) and dilute in 1 mL water.
6. Mix 1 µL of each of the two excised and diluted intermediate products from the first round of PCR in a second round of PCR of which the first cycle is run without primers. Half of the overlapping mutant strands act as primers on one another and generate mutant recombined fragments.
7. During the denaturation step of the second cycle, insert a micropipet tip through the mineral oil layer to add the biotinylated sense and the nonmodified antisense flanking primers in less than 1/10 the volume of the reaction. The recombined mutant fragments are amplified during subsequent cycles.
8. Separate biotinylated PCR products on a preparative agarose gel (*see* **steps 2–4**).
9. Purify fragment of interest using GeneClean.

3.3.2. Solid-Phase Synthesis and Magnetic Separation of Antisense Strands

Only anti-sense mutant fragments act as perfect competitors for first-strand cDNA *(2)*. A simple protocol for repeated synthesis and separation of the antisense strands from double-stranded biotinylated DNA fragments is listed below and schematically shown as part of **Fig. 1.**

1. Wash Streptavidin-coated magnetic beads (Dynabeads) with 1X binding and washing (B&W) buffer using a magnetic particle concentrator (MPC) for Eppendorf tubes.
2. Resuspend Dynabeads in 2X B&W buffer.
3. Mix equal volumes (50 μL) of the resuspended beads and the purified PCR. Use approx 0.3 mg Dynabeads per μg PCR product.
4. Incubate at room temperature for 30 min, keeping the beads suspended by gentle rotation of the tube.
5. Using the MPC, collect the immobilized DNA and remove the supernatant.
6. Wash beads extensively using the MPC with 3X 100 μL 1X B&W and 3X 100 μL TE to remove uncoupled PCR primers and products.
7. Resuspend beads in 20 μL of freshly prepared 0.1N NaOH solution and incubate at room temperature for 10 min to melt the DNA duplex.
8. Transfer alkali supernatant containing the nonbiotinylated antisense strands to a clean tube.
9. Neutralize supernatant with 4 μL of a freshly prepared 1:1 mixture of 1N HCl and 1M Tris-HCl (pH 8.0) (*see* **Note 6**).
10. Using the MPC, wash beads once with 0.1M NaOH (50 μL), once with 1X B&W (50 μL), and once with TE (50 μL) (*see* **Note 7**).
11. Resuspend beads in a prewarmed PCR mixture including all reaction components (*see* **Subheading 3.2.**) except for the sense primers.
12. Incubate for 10 min at a temperature between 55 and 60°C to allow both specific annealing and extension of the antisense primers (even below 55°C the polymerization rates of *Taq* are significant). Shake gently every 2 min or use a shaking water bath in order to keep the beads in suspension.
13. Using the MPC, collect beads and keep the supernatant containing the reaction mix at the desired annealing/extension temperature for successive rounds of solid-phase DNA synthesis. The immobilized sense strands can be reused several times for repeated solid-phase synthesis of antisense DNA.
14. Wash beads extensively using the MPC with 3X 100 μL 1X B&W and 3X 100 μL TE to remove free primers.
15. Again, resuspend and incubate beads in 20 μL 0.1N NaOH to separate the immobilized sense strands and the anti-sense strands (**steps 7–10**).
16. The procedure may be repeated several times to increase the yield of antisense strands: Wash beads, incubate beads in the PCR mixture (**steps 10–12**) that was kept at the desired annealing/extension temperature (**step 13**) and proceed as described in **steps 14** and **15**.
17. Dilute 1 μL neutralized anti-sense DNA in 24 μL EDTA/SDS.
18. Mix 10 μL of Formamide-gel-loading buffer and 10 μL of 25X diluted antisense DNA, denature at 95°C for 10 min and cool on ice.
19. Apply denatured antisense DNA on a 10% polyacrylamide gel and verify and quantify anti-sense competitors by electrophoresis under nondenaturing conditions and subsequent silver-staining and densitometric scanning using a concentration standard (*see* **Note 8**).

20. Make a serial dilution of antisense competitors down to a concentration of 0.1 fg/μL and add carrier DNA to a final concentration of 1 ng/μL to each competitor concentration (*see* **Note 9**).
21. Heat-denature these DNA samples at 94°C for 5 min.
22. Cool down slowly to hybridize the single-stranded competitor DNA to denatured carrier DNA thus stabilizing the DNA in double-stranded structures.
23. Aliquotize and store at –20°C in siliconized tubes.

3.4. CD-RT-PCR and Restriction Enzyme Digestion

This section describes how to perform CD-RT-PCR up to the quantification of the digested reaction products. Before doing so, it is helpful to gain some information about the primer concentrations that have to be applied for successful co-amplification of the gene of interest and the reference gene and about the approximate amount of competitor DNAs that have to be used for detectable competition with both cDNAs. Hence, it is recommended to perform a few simple pilot experiments (*see* **Note 10**) in order to minimize the number of reactions required for gene expression analysis of a single sample. In many cases a single reaction will do since quantification by CD-RT-PCR is accurate within a wide range of different competitor concentrations. After tuning the CD-RT-PCR in a few pilot experiments, measurement in samples with a 35-fold difference in expression levels can be performed with a single primer and competitor concentration *(2)*. PCR is carried out basically as described in **Subheading 3.2.** The protocol below describes a CD-RT-PCR for a series of samples with a wider range of expression levels (>35-fold) in which a single competitor concentration is expected to yield inconclusive results (*see* **Note 11**).

1. Set up a PCR master mix with a fixed amount of cDNA and other components for separate reactions with different competitor concentrations. Primers for both the gene of interest and the reference gene are added to this mixture. Leave out the volume necessary to add competitors to the reactions (*see* **Note 12**).
2. Divide the master mix into the different reactions.
3. Add exactly defined amounts of antisense mutant competitor DNA of both the gene of interest and the reference gene to the different reactions. It is recommended to choose competitor dilutions with a 100-fold difference in concentration and add equal volumes of these dilutions to the different reactions (*see* **Notes 12** and **13**).
4. Run PCRs.
5. Heat-denature amplification products at 94°C for 3 min.
6. Cool down slowly to room temperature within 1.5–2 hours to promote stabilization of heteroduplexes based on random re-annealing of nearly homologous mutated and non-mutated sequences (*see* **Note 14**).
7. Digest DNA (*see* **Note 15**) without precipitating it by adding the appropriate restriction enzyme in a mix with KGB to a portion (5 to 10 μL) of the PCR (*see* **Note 16**).

8. To assess the success of the reactions, load a few samples on an ethidium bromide stained gel system. Use high percentage (6%) agarose gels (NuSieve 5:1 agarose) if the samples were digested in KGB since the buffer influences electrophoresis properties negatively. A pattern of four bands should appear representing the undigested fragments and the overlapping halves of digested fragments of the two different genes and their corresponding competitors.

3.5. HPLC Analysis of CD-RT-PCR Products After Restriction Enzyme Digestion

A trustworthy estimation of the ratios of the DNA products after CD-RT-PCR and restriction enzyme digestion relies on a technique that offers efficient and quantitative discrimination of the resulting four different fragment sizes. HPLC analysis of DNA comprises direct UV absorbance measurement resulting in a high linearity over a large range of DNA concentrations *(7)*. For this reason and because of the sensitivity of the technique, HPLC is highly suitable for measurement of CD-RT-PCR products.

1. Prepare HPLC system including the analytical column as was also described by Katz in a previous edition of the Methods in Molecular Biology series *(8)*.
2. Program a gradient profile. For a fast and efficient separation of 50–300 bp fragments that allows the injection of consecutive samples every 15 min, a linear mobile phase change of buffer B in A is recommended as follows:
   a. From 70 to 60% B in 10 s
   b. From 60 to 48% B in 3 min
   c. From 48 to 42% B in 5 min
   d. From 42 to 0% B in 10 s, and hold for 1 min for clean-up
   e. From 0 to 70% in 10 s, and hold for 5 min for reequilibration
3. Inject the digested CD-RT-PCR samples and run the gradient profile.
4. Verify whether the retention times of the appearing peaks are as expected by comparison with the retention times of fragments of an appropriate standard that is injected periodically.
5. Collect integrated data of the area under the curve of relevant peaks.

3.6. Calculation of Normalized Expression Levels

1. Calculate the proportion of digested fragments *(d)* for both genes separately from the area under the curve (AUC) of the peaks that correspond to the digested (dig) and undigested (undig) amplification products of a gene:

$$d = AUC_{dig}/(AUC_{dig} + AUC_{undig}) \qquad (1)$$

2. Calculate the amount of specific cDNA *(Y)* prior to PCR for both genes separately. The initial amount of competitor *(C)* for each cDNA species that was added to the reaction is known. Fill in *C* and *d* into the following equation (*see* **Note 17**), where n = +1 or –1 depending on whether the competitor possesses (+) or lacks (–) the unique restriction enzyme recognition sequence:

$$Y = C[(1-\sqrt{d})/\sqrt{d}]^n \qquad (2)$$

3. Present the expression level (X) of the gene of interest (i) in a molar relation to the internal standard or reference gene (r) by inclusion of the size (S) of the amplified fragments (in basepairs) into the following equation:

$$X_i = (Y^i/Y_r).(S_r/S_i) \qquad (3)$$

4. Notes

1. Typical primers have a GC content as close to 50% as possible. Primers used for amplification of cDNA and competitor DNA during CD-RT-PCR were 20 to 22 bases in length. Pairs of mutagenic primers are recommended to be 30 to 35 bases in length, with base substitutions (one to four) centrally located in a 3' region of 17 to 20 bases in length that confers opposite primers complementary to one another (*see also* **Fig. 1**). To avoid problems due to amplification of contaminating genomic DNA, the sense and antisense primers for CD-RT-PCR should reside in different exons.

   The position of CD-RT-PCR primers should also be carefully considered with respect to the size of the fragments to be amplified. CD-RT-PCR analysis requires adequate measurement of both the full-size amplification products of two genes (A and B) and the fragments that result from cutting these products in two equal halves. To achieve equal intervening distances between the resulting four bands on gel or HPLC peaks (*see* **Fig. 2**), the size (S) of gene A and B should relate as follows: $S_A = \sqrt{2} \cdot S_B$. It is recommended to choose 200 bp $\leq S_A \leq$ 300 bp. S_A will then be small enough for amplification with randomly primed cDNA as the template and $S_B/2$ will not be too small for gel electrophoresis and HPLC.

   Primer design and the choice of an appropriate restriction enzyme for CD-RT-PCR are mutually dependent. Mutagenic primers may be used to create or remove a unique restriction enzyme site within the sequence region that is to be amplified during CD-RT-PCR. Choose these primers in a way that the unique restriction enzyme sites reside exactly in the middle of the amplification products of CD-RT-PCR (*see also* **Fig. 2**). The two halves of digested fragments will then elute at (about) the same time during HPLC analysis, which lowers the detection limit due to increased AUCs and reduces the complexity of the chromatograms.

2. It is not required to know the RNA concentration of the samples as long as quantifiable amounts of specific DNA can be amplified from it. CD-RT-PCR may therefore be applied to measure RNA expression on even minute amounts of mRNA from biopsies, needle aspirations, and so on.

3. In CD-RT-PCR, expression levels are presented as cDNA ratios of the target and the reference gene. Linearity between these ratios and normalized mRNA expression levels is achieved by using randomly annealing hexamer primers in reverse transcription. In contrast to oligo(dT) or gene specific oligo priming, randomly primed reverse transcription results in equally efficient cDNA synthesis of different RNA species and overcomes problems due to sequence complexity and mRNA secondary structure.

4. Specificity and sensitivity may be further improved by performing a hot-start PCR *(9)* and/or a nested primer strategy. It is recommended for both RT reactions and PCRs to prepare a master mix of all invariable components of a series of reactions and add aliquots of this mix to the variable reaction components in each separate reaction tube.

5. Beware of UV-radiation-induced DNA mutagenesis. Work as quickly as possible and/or use a low-energy UV trans-illuminator.

6. Perform a titration of HCl(1:1)Tris-HCl against NaOH before using the solutions for neutralization of the DNA samples. Always use the same pipet for both solutions. Calibration differences between different pipets may cause neutralization problems.

7. The immobilized sense strands can be stored in TE at 4°C for several weeks. Before using the sense DNA coupled to the beads as template for solid-phase synthesis, start with alkali denaturation (**Subheading 3.3.2., step 7**)

8. The antisense DNA can easily be distinguished from the sense strands based on the principle that these complementary single-stranded DNA fragments have different sequence-dependent conformations that influence electrophoretic mobility *(5)*. As a consequence, denatured PCR products appear as two separate bands on a nondenaturing gel. After the magnetic separation procedure, only the band that represents the antisense strands remains *(2)*. It is recommended to set aside antisense DNA from the first time of alkali denaturation and pool DNA from subsequent cycles of solid-phase synthesis and denaturation since the purity of antisense DNA from the first harvesting is often not as required.

9. Use a single well-calibrated pipette for the preparation of a competitor DNA dilution series to avoid concentrations errors. We used a Gilson P200 pipet and combined 50 µL competitor DNA (from a 10X concentration), 50 µL carrier DNA (10 ng/µL) and 2X 200 µL water.

10. First, to find the conditions that yield quantifiable amounts of PCR products both for the gene of interest and the reference gene, perform a series of differential RT-PCRs (without competitors) by adding the primers for these genes in different concentration ratios to the reaction mixtures. Second, titrate the competitors against a fixed amount of cDNA. The four bands/peaks comprising the digestible and nondigestible DNA fragments of both the gene of interest and the reference gene should all be present in quantifiable amounts after CD-RT-PCR and restriction enzyme digestion (*see* **Fig. 2**).

11. To cover extremely wide ranges of expression levels it may sometimes be necessary to also use two different primer concentration ratios.

12. Use a single, well-calibrated pipet for the preparation of a CD-RT-PCR to reduce experimental variation. To ensure adequate pipeting, the volumes should not be too low. A 25 µL CD-RT-PCR could be made up by 15 µL of a master mix that was divided into separate reactions and 10 µL of competitors.

13. The reference gene can be fine-tuned more precisely when the reactions are roughly standardized with respect to the input of cDNA amounts by taking dilutions with a 10-fold difference in concentration.

14. Most PCR machines permit the programming of a temperature ramping rate to slowly cool the samples. Alternatively, a water bath may be used.

15. As a control for restriction enzyme digestion also digest a PCR that solely contains fully digestible DNA. One could also add a DNA fragment carrying the restriction enzyme site as an internal control to every sample after denaturation and re-annealing.

16. KGB should be diluted to a final reaction concentration varying between 0.5 and 2.0X KGB depending on the optimal reaction conditions for the restriction enzyme used. Digestion with EcoRI was performed in 0.5X KGB: 7 μL PCR + 5 μL enzyme mix containing 3 μL H_2O + 1 μL 6X KGB + 1 μL EcoRI.

17. The equation is based on the quadratic distribution principle of duplex formation. **Figure 2** shows an example where a one to one ratio of first-strand cDNA to competitor DNA before PCR renders one out of four duplexes distinguishable from the others by restriction enzyme digestion.

References

1. Rochlitz, C. F., de Kant, E., Neubauer, A., Heide, I., Bohmer, R., Oertel, J., Huhn, D., and Herrmann, R. (1992) PCR-determined expression of the MDR1 gene in chronic lymphocytic leukemia. *Ann. Hematol.* **65**, 241–246.

2. de Kant, E., Rochlitz, C. F., and Herrmann, R. (1994) Gene expression analysis by a competitive and differential PCR with antisense competitors. *BioTechniques* **17**, 934–942.

3. Higuchi, R., Krummel, B., and Saiki, R. K. (1988) A general method of in vitro preparation and specific mutagenesis of DNA fragments: study of protein and DNA interactions. *Nucleic Acids Res.* **16**, 7351–7367.

4. Maniatis, T., Fritsch, E. F., and Stambrook, J. (1989). *Molecular Cloning. A Laboratory Manual (Second Edition).* Cold Spring Harbor Laboratory Press, Cold Spring Harbor, New York.

5. Orita, M., Suzuki, Y., Sekiya, T., and Hayashi, K. (1989) Rapid and sensitive detection of point mutations and DNA polymorphisms using the polymerase chain reaction. *Genomics* **5**, 874–879.

6. Chomzynski, P. and Sacchi, N. (1987) Single-step method of RNA isolation by acid guanidinium thiocyanate-phenol-chloroform extraction. *Anal. Biochem.* *162*, 156–159.

7. Katz, E. D. and Dong, M. W. (1990) Rapid analysis and purification of polymerase chain reaction products by high-performance liquid chromatography. *BioTechniques* **8**, 546–555.

8. Katz, E. D. (1993) Quantitation and purification of polymerase chain reaction products by high-performance liquid chromatography in *Methods in Molecular Biology: PCR Protocols: Current Methods and Applications*, vol. 15, (White, B.A., Ed.) pp.63–74. Humana, Totowa, NJ.

9. Chou, Q., Russell, M., Birch, D. E., Raymond, J., and Bloch, W. (1992) Prevention of pre-PCR mis-priming and primer dimerization improves low-copy-number amplifications. *Nucleic Acids Res.* **20**, 1713–1723.

26

Primer Extension Analysis of mRNA

Maggie Walmsley, Mark Leonard, and Roger Patient

1. Introduction

Primer extension is a relatively quick and convenient means by which gene transcription can be monitored. The technique can be used to determine accurately the site of transcription initiation or to quantify the amount of cap site-specific message produced.

The principle of this technique is outlined in **Fig. 1.** In brief, a radiolabeled primer fragment (usually a single-stranded oligonucleotide of approx 20 nucleotides long) is hybridized to its complementary sequence near the mRNA 5' terminus. The primer is then extended by the enzyme reverse transcriptase back to the initiation point (cap site) of the message. The products of the reaction are run out on a denaturing polyacrylamide gel and exposed to autoradiography.

The major advantage of primer extension for RNA analysis is its convenience (compared to S1 mapping or RNase protection). This technique enables precise determination of the start point of transcription of a newly isolated gene. Also, because very clean results can be obtained, more than one mRNA can be quantitatively analyzed in a single reaction. For example, the transcripts from a transfected marked gene can be distinguished in size from the endogenous gene, or, by the judicious choice of primers, transcription from a test gene and a cotransfected (internal standard) control gene can be monitored in the same reaction (*see* **Fig. 2**). Finally, for genes with multiple cap sites, the amount of transcription from each site can be distinguished in a single reaction.

Many of the above considerations apply equally to the analysis of mRNA by S1 or RNase mapping. Although RNase protection is usually more sensitive, the convenience of primer extension means that it is often the method of choice for RNA analysis. In the situation in which the 5' end of a new gene is being determined, however, it is important to perform both primer extension and

From: *Methods in Molecular Biology, Vol. 86: RNA Isolation and Characterization Protocols*
Edited by: R. Rapley and D. L. Manning © Humana Press Inc., Totowa, NJ

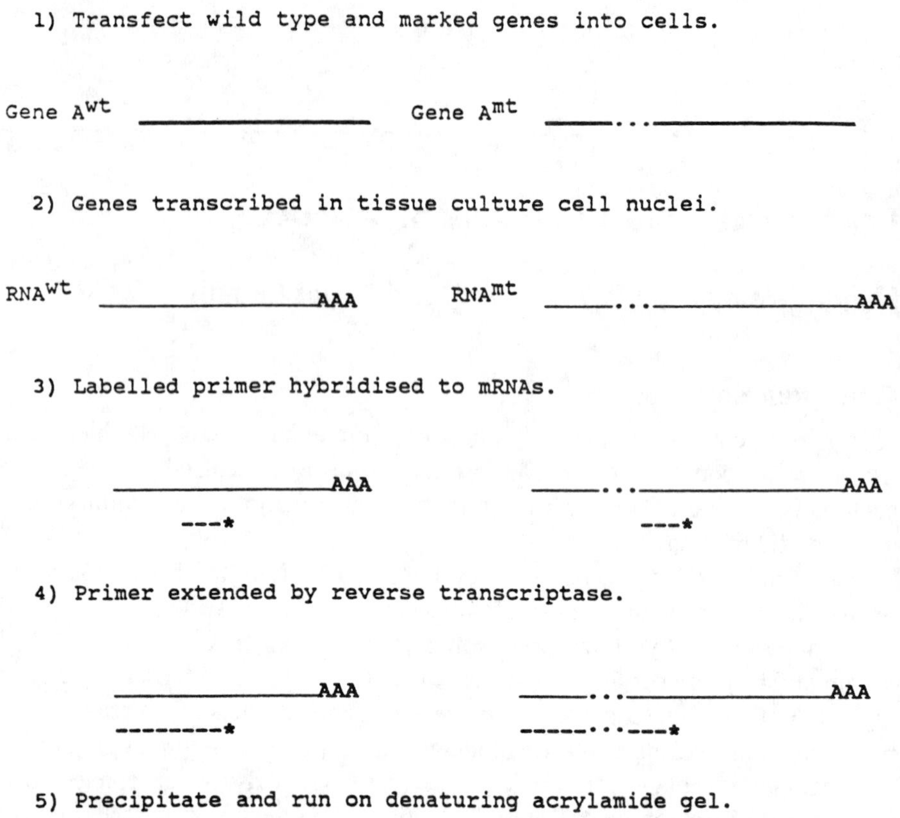

```
1) Transfect wild type and marked genes into cells.
```

Gene A^{wt} _____   Gene A^{mt} _____...._____

```
2) Genes transcribed in tissue culture cell nuclei.
```

RNA^{wt} _____AAA   RNA^{mt} _____...._____AAA

```
3) Labelled primer hybridised to mRNAs.
```

_____AAA   _____...._____AAA
      ---*   ---*

```
4) Primer extended by reverse transcriptase.
```

_____AAA   _____...._____AAA
   --------*   -----···---*

```
5) Precipitate and run on denaturing acrylamide gel.
```

Fig. 1. Distinguishing the different sized transcripts from wild-type and marked copies of the *X. laevis* β-globin gene by primer extension: schematic representation of the primer extension reaction.

either S1 or RNase mapping to identify the cap site unambiguously because all these methods generate artifacts that can result in misassignment of the 5' end. Premature termination of synthesis, perhaps due to RNA secondary structure, leads to less than full-length products in primer extension. Internal cleavage of nucleic acid hybrids, possibly at AT-rich tracts that undergo local denaturation, leads to foreshortened products in S1 and RNase mapping. Because the artifactual bands generated have different causes, it is normally safe to assume that the correct start site has been identified when both methods identify the same cap site.

2. Materials

As with all procedures involving RNA, extreme care must be taken to avoid RNase contamination and degradation of samples *(1)*. In recent years, the

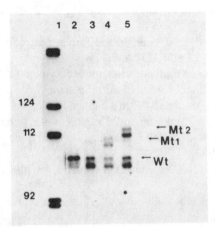

Fig. 2. An example of a primer extension reaction: total cytoplasmic RNA from cultured cells containing *Xenopus* β-globin gene constructs, analyzed using a single 5'-labeled oligo primer. Lane 3, wild-type, β-globin gene introduced alone; Lanes 4 and 5, wild-type gene plus genes marked by insertion of different-sized oligonucleotides into the first exon. Samples are run alongside a labeled DNA size marker (Lane 1) and total cytoplasmic RNA (50 ng) from *X. laevis* erythroblasts (Lane 2) to show up the cap site used in vivo.

advent of ribonuclease-inhibiting enzymes, such as RNasin, have rendered the use of the RNase inhibitor diethylpyrocarbonate (DEPC) redundant, assuming reasonable care is taken to work cleanly. Work surfaces should be clean and gloves should be worn at all times. Stock solutions, glassware, and plastic pipet tips should be autoclaved. Whenever possible use sterile plasticware in place of glass. Separate stocks of reagents and plasticware should be maintained solely for RNA work.

The most convenient source of single-stranded primer is a chemically synthesized short oligonucleotide that also yields a labeled probe of high specific activity (*see* **Note 1**). The primer should be 18 nucleotides or longer, lacking a terminal phosphate at the 5' end to facilitate labeling with polynucleotide kinase, and should ideally be located within 100–200 base pairs of the 5' end of the mRNA. Care should be taken that the chosen oligomer does not contain repeats capable of inter- or intraprobe hybridization, since this will reduce the efficiency of annealing to mRNA.

1. 10X Kinase buffer: 500 mM Tris-HCl, pH 7.6, 100 mM MgCl$_2$, 50 mM dithiothreitol (DTT), 1 mM spermidine, 1 mM EDTA. Alternatively use supplier's buffer.
2. 5X Hybridization buffer: 2M NaCl, 50 mM PIPES, pH6.4 (autoclave before use).
3. 1X Extension buffer: 10 μL of 1M Tris-HCl, pH 8.3, 10 μL of 200 mM DTT, 10 μL of 120 mM MgCl$_2$, 5 μL of 1 mg/mL actinomycin D (care, toxic and light

sensitive), 10 µL each of 10 mM dATP, dCTP, dGTP, dTTP, 1 µL (40U) RNasin and water to 178 µL.

4. Formamide loading dye: 80% deionized formamide, 45 mM Tris-HCl, pH 8.3, 45 mM boric acid, 1.25 mM EDTA, 0.02% xylene cyanol.

5. G25 Sephadex: Suspend Sephadex in approx 50 vol of TE (10 mM Tris-HCl, pH 8.0,1 mM EDTA). Autoclave for 15 min (Sephadex will swell) and allow to cool to room temperature. Store at 4°C in a capped bottle.

6. Carrier (yeast) tRNA: Make up to 25 mg/mL with autoclaved water. Use the maker's bottle to avoid RNase contamination during weighing. Store in sterile plastic tubes at –20°C.

7. 3M sodium acetate (NaOAc), brought to pH 5.4 with glacial acetic acid and autoclaved for 15 min.

8. 70% Ethanol: 35 mL absolute ethanol, 15 mL autoclaved water made up in a sterile 50 mL plastic Falcon tube.

3. Methods
3.1. Kinasing the Primer

1. In an autoclaved microfuge tube, mix 1–5 pmol single stranded oligonucleotide (*see* **Note 1**), 1 µL 10X kinase buffer, 35 µCi γ^{32}P ATP (3000 Ci/mmol), 20U T4 polynucleotide kinase, and autoclaved water to 10 µL.

2. Incubate at 37°C for 30 min.

3. Add 2 µL of 250 mM EDTA.

4. Remove unincorporated labeled nucleotide by centrifugation down a Sephadex G25 column (*see* **Note 2**).

5. Transfer the eluate (which contains the probe) to an autoclaved microfuge tube and heat-kill the kinase by incubating at 65°C for 10 min. Extract with an equal volume of phenol:chloroform:isoamyl alcohol (25:24:1) (v/v).

6. To the aqueous phase add sodium acetate and MgCl$_2$ to a final concentration of 0.3M and 5 mM respectively and add 10 µg tRNA.

7. Ethanol precipitate by adding 2.5 vol of 100% ethanol, placing in dry ice for 10 min, and spin at high speed in a microfuge for 30 min at 4°C.

8. Wash the pellet in 70% ethanol, dry under vacuum (approx 2 min) and resuspend at 1–10 fmol/µL in autoclaved water.

3.2. Hybridization Reaction

1. In an autoclaved microfuge tube, mix 1–50 fmol-labeled primer (*see* **Note 3**), 4 µL 5X hybridization buffer, 0.1–5 fmol target mRNA (up to 20 µg total RNA, *see* **Note 4**) and autoclaved water to a final volume of 20 µL.

2. Heat the samples to 70°C for 3 min.

3. Hybridize for 1.5–4h (*see* **Note 4**) in a water bath at approx 54°C or at an empirically determined optimum temperature (*see* **Note 5**).

3.3. Extension Reaction

1. To each sample add 178 µL extension buffer and place the tube on ice.

2. Add 1 µL RNasin and 1 µL reverse transcriptase per sample.

3. Transfer tubes to 42°C and incubate for 1 h (*see* **Note 6**).
4. Add 20 μL 3*M* NaOAc, and precipitate with 2.5 vol EtOH. Transfer to dry ice for 10 min and spin at high speed in a microfuge for 15 min.
5. Wash with 70% EtOH and vacuum dry for 2 min.
6. Take pellet up in 10 μL formamide dyes, being sure to resuspend thoroughly.
7. Heat denature at 90°C for 3 min and chill on ice.
8. Load 8 μL on an appropriate percentage denaturing polyacrylamide gel and electrophorese along with suitable size markers or DNA sequencing ladders.

The labeled extension products are run on a standard denaturing urea-acrylamide gel of suitable percentage (dependent on the distance of the primer from the 5' end of the message) to enable good resolution of cap site length transcripts. This is particularly important when the site of transcription initiation is to be determined, but is also important for quantative experiments if multiple start sites exist in the gene of interest (*see* **Notes 7–9**).

4. Notes

1. The sensitivity of the primer extension analysis can be improved by using a uniformly labeled M13 vector-generated probe. However, except for the detection of extremely rare transcripts, this should not be necessary. To generate a probe of maximal specific activity using a kinased oligomer, label no more than 5 pmoles of a single-stranded oligomer (enough for 500 assays) at a time. In our experience, the kinasing reaction is not linear above this point and labeling larger amounts is counterproductive, leading to a lowering of specific activity.
2. Prepare G25 Sephadex as follows: Plug the nipple of a 2-mL sterile plastic syringe with polymer wool and fill the syringe with a concentrated slurry of Sephadex. Suspend the syringe over a sterile plastic centrifuge tube and spin at 1500 rpm in a swing-out rotor for 5 min. Load exactly 100 μL TE on to the column and spin again for 5 min at 1500 rpm. Discard the centrifuge tube and buffer contents, and transfer the syringe to a fresh tube. Load the labeled primer made up to 100 μL with TE and spin for 5 min at 1500 rpm. Unincorporated label will be located at the top of the column and labeled primer will be found in the eluate which should measure close to 100 μL. If the eluate measures significantly less than 100 μL, wash the column with a further 50–100 μL, centrifuge as before and combine eluates. Remove 1 μL of the probe, add to 100 μL water and determine the specific activity (SA) by Cerenkov counting. Probes should have an SA of 2–5 × 10^6 cpm/pmol.
3. Alternatively, probe and test RNA can be coprecipitated with ethanol. Take up the pellet in autoclaved water and add the remaining components of the hybridization mix.
4. The duration of the hybridization reaction is dependent on the amount of probe and target present in the reaction mix. In the case of transfected or injected cells in which transcripts are abundant, we are dealing with the top end of the range of probe and target concentrations and hybridization over 90 min (or less—the minimum time may be determined empirically) will suffice to drive the reaction to completion. When probing for low abundance endog-

enous transcripts, however, less target and probe are being used and hybridization times should be increased to a minimum of 4 h. The hybridization is carried out with the primer in approx 10-fold molar excess over the target RNA. Too great an excess (particularly at suboptimal hybridization stringencies) can result in nonspecific priming. In cases in which it is not possible to estimate the abundance of the target mRNA, the following procedure may be followed to ensure that the hybridization is being carried out in primer excess, which is necessary if quantitative results are required. Include three or four extra samples, which represent a small titration of an RNA stock that contains the target mRNA (over a range of 1–20-fold, for example). If the signals from your test samples fall within the linear range of the titration, then the assay has been carried out in primer excess.

5. The optimum temperature for hybridization will depend on the length of primer and its base composition. Formulae for the estimation of RNA-DNA hybrid melting temperatures *(2,3)* are not accurate for short DNA primers. For most probes the value is in the range of 45–65°C in the buffer given, but if maximum sensitivity is required, pilot experiments should be carried out over this range to determine the optimum temperature for the specific primer–mRNA combination being used.

6. The elongation reaction is carried out at 42°C to reduce the amount of RNA secondary structure, which can result in premature termination of the extension reaction.

7. The further the primer is located from the cap site, the more likely the occurrence of premature termination by reverse transcriptase at sites of RNA secondary structure. The presence of discrete "drop off" bands will result in a decreased cap site signal. Changing the position of the primer to a site 5' of secondary structure barriers will improve the cap site signal.

8. The extension reaction often generates more than one band in the vicinity of the cap site (*see* **Fig. 2**). These may represent genuine multiple starts of transcription. In addition, methylation of the mRNA at the cap site, or adjacent nucleotide, may cause premature termination in a number of molecules. The ratio of cap site bands can vary between different sources of the mRNA but is constant for a given source.

9. Additional bands may be produced as a result of fold-back cDNA synthesis by reverse transcriptase. The sequence of the 5' end of a new message should be examined closely for the possibility of such fold-back structures. Spurious bands may arise by cross hybridization of the primer to endogenous tissue culture cell RNA (for example) or carrier tRNA. Negative controls using RNA from nontransfected culture cells and tRNA will identify such bands.

References

1. Blumberg, D. (1987) Creating a ribonuclease-free environment, in *Methods in Enzymology*, vol. 152 (Berger, S. L. and Kimmel, A. R., eds.), Academic, New York. pp. 20–24.

2. Thomas, M., White, R. L., and Davis, R. W. (1976) Hybridization of RNA to double-stranded DNA. *Proc. Nat. Acad. Sci. USA* **73**, 2294–2298.
3. Britten, R. J. and Davidson, E. H. (1985) Hybridisation strategy, in *Nucleic Acid Hybridisation–A Practical Approach* (Hames, B. D. and Higgins, S. J., eds.), IRL, Oxford, Washington, DC, pp. 3–15.

Suggested Readings

Calzone, F. J., Britten, R., and Davidson, E. H. (1987) Mapping gene transcripts by nuclease protection assays and cDNA primer extension, in *Methods in Enzymology*, vol. 152 (Berger, S. L. and Kimmel, A. R., eds.), Academic, New York, pp. 611–632.
Krug, M. S. and Berger, S. L. (1987) First strand cDNA synthesis primed with oligo dT, in *Methods in Enzymology*, vol. 152 (Berger, S.L. and Kimmel, A.R., eds.), Academic, London, pp. 316–325
Williams, J. G. and Mason, P. J. (1985) Hybridization and analysis of RNA, in *Nucleic Acid Hybridisation—A Practical Approach* (Hames, B. D. and Higgins, S. J., eds.), IRL, Oxford, Washington, DC, pp. 139–160.

27

S1 Mapping Using
Single-Stranded DNA Probes

Stéphane Viville and Roberto Mantovani

1. Introduction

The S1 nuclease is an endonuclease isolated from *Aspergillus oryzae* that digests single- but not double-stranded nucleic acid. In addition, it digests partially mismatched double-stranded molecules with such sensitivity that even a single base-pair mismatch can be cut and hence detected. In practice, a probe of end-labeled double-stranded DNA is denatured and hybridized to complementary RNA molecules. S1 is used to recognize and cut mismatches or unannealed regions and the products are analyzed on a denaturing polyacrylamide gel. A number of different uses of the S1 nuclease have been developed to analyze mRNA taking advantage of this property *(1,2)*. Both qualitative and quantitative information can be obtained in the same experiment *(3)*.

Qualitatively, it is possible to characterize the start site(s) of mRNA, to establish the exact intron/exon map of a given gene (*see* **ref.** *4* and **Fig. 1**), and to map the polyadenylation sites. Quantitatively, it can be used to study gene regulation both in vivo and in vitro, for example, in the study of the Eα gene promoter *(5)*.

In this chapter we describe a S1 mapping method based on the preparation and use of a single-stranded DNA probe (*see* **Fig. 2**). This offers many advantages:

1. Oligonucleotides allow the exact choice of fragment for a probe.
2. Oligonucleotide labeling is easy and efficient, resulting in a high specific activity probe.
3. The probe can be prepared from single-stranded DNA (e.g., M13, BlueScript) as well as a double-stranded template.
4. The single-stranded probe avoids problems often encountered setting up hybridization conditions of double-stranded probes *(6)*.
5. The probe is stable for 3–4 wk.

We illustrate the process with examples from the analysis of the Eα promoter.

From: *Methods in Molecular Biology, Vol. 86: RNA Isolation and Characterization Protocols*
Edited by: R. Rapley and D. L. Manning © Humana Press Inc., Totowa, NJ

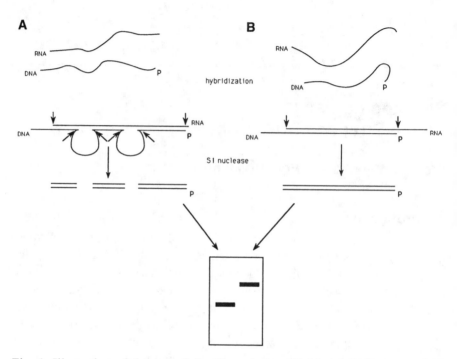

Fig. 1. Illustration of the use of the S1 nuclease. (**A**) Mapping the intron/exon organization of a given gene. The labeled DNA hybridizes to the corresponding mRNA; the introns are cut by the S1 nuclease. (**B**) Mapping of the 5' end of mRNA. The labeled DNA is used as a probe to map the 5' end of a messenger RNA. In both cases the labeled fragments are visualized on a denaturing acrylamide gel.

2. Materials

2.1. Preparation of Single-Stranded DNA Probe from a Single-Stranded DNA Template

1. 10X Kinase buffer: 400 mM Tris-HCl, pH 7.8, 100 mM MgCl$_2$, 100 mM β-mercaptoethanol, 250 μg/mL bovine serum albumin. Store at −20°C.
2. [^{32}P]-γ-ATP: specific activity >3000 Ci/mmol.
3. Suitable oligonucleotide (*see* **Note 1**). Prepare a 10 pmol/μL solution in distilled water. Store at −20°C.
4. PNK: Polynucleotide kinase at a stock concentration of 10 U/μL. Store at −20°C.
5. 10X Annealing buffer: 100 mM Tris-HCl, pH 7.5, 100 mM MgCl$_2$, 500 mM NaCl, 100 mM dithiothreitol.
6. DNA template: 1 mg/mL (*see* **Note 2**).
7. 10X dNTP mix: 5 mM dATP, 5 mM dCTP, 5 mM dGTP, and 5 mM dTTP.
8. Klenow: The large fragment of *E. coli* DNA Polymerase I at a stock concentration of 10 U/μL. Store at −20°C.

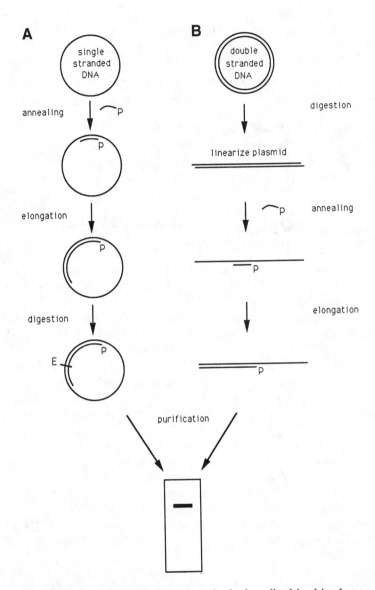

Fig. 2. Schematic illustration of the two methods described in this chapter for the preparation of a single-stranded DNA probe. **(A)** Using a single-stranded plasmid as a template. **(B)** Using a double-stranded DNA plasmid to synthesize the probe.

9. Suitable restriction enzyme (*see* **Note 3**).
10. 6% Polyacrylamide/8M urea solution in 0.5X TBE and formamide dye.
11. X-ray film: A film of suitable sensitivity, e.g., Kodak XAR.
12. Elution buffer: 50 mM Tris-HCl, pH 7.5, 0.5 mM EDTA.

13. 3*M* sodium acetate, pH 7.4.
14. tRNA: Stock solution 10 mg/mL. Store at –20°C.

2.2. Hybridization and S1 Analysis

15. 4X Hybridization buffer: 1.6*M* NaCl, 40 m*M* PIPES, pH 6.4.
16. Deionized formamide.
17. Paraffin oil.
18. S1 buffer: 300 m*M* NaCl, 30 m*M* sodium acetate, pH 4.5, 4.5 m*M* zinc acetate.
19. S1 enzyme: A stock concentration of 10 U/µL. Store at –20°C.
20. S1 stop buffer: 2.5*M* ammonium acetate, 50 m*M* EDTA.
21. Isopropanol.

3. Methods

3.1. Preparation of Single-Stranded DNA Probe from a Single-Stranded DNA Template

1. Label the oligonucleotide by incubation of 1 µL of cold oligonucleotide, 10 µL of [^{32}P]-g-ATP, 2 µL of 10X kinase buffer, 6 µL of dH$_2$O, and 1 µL of PNK at 37°C for 45 min. Inactivate the PNK at 95°C for 2 min.
2. Add 2 µL of single-stranded DNA template, 4 µL of 10X annealing buffer, and 14 µL of H$_2$O and incubate for 10 min at 65°C, followed by 10 min at 55°C, then 10 min at 37°C.
3. Add 4 µL of dNTP mix and 1 µL of Klenow and leave at room temperature for 10 min. Inactivate the Klenow by placing the tube at 65°C for 15 min.
4. Correct the mix for the restriction enzyme digest by either adding a suitable amount of sodium chloride or by dilution. Add 20 U of restriction enzyme and incubate at 37°C for 60 min (*see* **Note 3**).
5. Precipitate the sample by adding 2 µL of tRNA stock, 0.1 vol of 3*M* sodium acetate, and 3 vol of ethanol, and place the sample in dry ice for 10 min. Centrifuge at 12,000*g* for 15 min in a microfuge.
6. Resuspend the pellet in 20 µL of formamide dye, heat at 95°C for 5 min, and load a 6% polyacrylamide/urea gel. Run the gel until the bromophenol blue reaches the bottom.
7. Locate the single-stranded DNA fragment by exposing the gel for 5 min to a X-ray film. Excise the corresponding band and elute it overnight in 500 µL of elution buffer.
8. Add 50 µL of sodium acetate, 20 µg of tRNA, and 1 mL of ethanol; place at –20°C for 2 h. Centrifuge for 20 min at 12,000*g* in a microfuge.
9. Count the pellet and resuspend it in formamide (10^5 cpm/µL). Store at –20°C. The probe is now ready for hybridization and S1 analysis.

3.2. Preparation of Single-Stranded DNA Probe from Double-Stranded DNA Template

Follow the same protocol described in **Subheading 3.1.**, but for the following exceptions (*see* **Fig. 2B**).

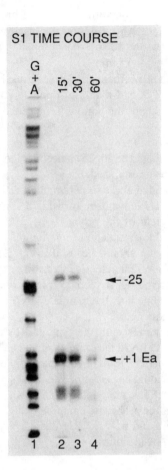

Fig. 3. Time-course of S1 analysis. In vitro synthesized RNA hybridized to Eα probe is incubated at 37°C for the indicated time with 100 U of S1 nuclease.

1. Initially cut 20 μg of double-stranded plasmid with a restriction enzyme of choice. Precipitate and resuspend in H_2O at a concentration of 1 mg/mL.
2. The annealing step (**Subheading 3.1., step 2**) must be carried out extremely quickly: After heat inactivation of PNK, add 2–5 μg of restricted plasmid, 4 μL of 10X annealing buffer, and 14 μL of H_2O, and incubate for 5 min at 95°C. Immediately place the tube in ice-cold water and then proceed to step 3.
3. Omit the restriction enzyme digestion (**Subheading 3.1., step 4**).

3.3. Hybridization and S1 Analysis (see Fig. 3)

1. In a 0.5 mL Eppendorf tube put 10 μL of probe (10,000 cpm total), 5 μL of 4X hybridization buffer, and 5 μL of RNA (total poly A$^+$ or synthesized in vitro). Add 20 μL of paraffin oil to prevent evaporation and incubate 4–16 h at 37°C (*see* **Note 4**).

2. Add 200 μL of S1 buffer containing 100 U of S1 nuclease, mix well, and place at 37°C for 5–30 min (*see* **Note 5**). Stop the reaction by adding 50 μL of S1 stop buffer, 40 μg of tRNA, and 300 μL of isopropanol. Place in dry ice for 15 min and centrifuge at 12,000g for 15 min.

3. Wash the pellet with 80% ethanol and dry. Resuspend in 3 μL of formamide dye, heat for 5 min at 95°C, and load a 6% polyacrylamide/urea gel. Run until the bromophenol blue reaches the bottom. Expose the gel for 8–48 h (*see* **Fig. 3**).

4. Notes

1. The oligonucleotide should be 20–25 nucleotides long and should be complementary to the DNA strand to be analyzed.

2. Both the single-stranded and double-stranded DNAs are prepared according to standard protocols. The advantage of using a single-stranded DNA template is basically a higher yield of the probe: 1–1.5×10^6 cpm vs the 4–6×10^5 cpm expected from a double-stranded plasmid template.

3. The restriction enzyme should be used to cut 200–600 nucleotides 3' of the position to which the oligonucleotide hybridizes to increase fragment recovery. Longer fragments are harder to recover from acrylamide gels. It does not matter if the restriction enzyme cuts the template DNA more than once as long as it is outside the probe sequence.

4. Hybridization time varies with the type of RNA to be analyzed: Using total cytoplasmic or poly A$^+$ RNA, the samples should be left at 37°C at least 12 h; when hybridizing RNA generated from in vitro transcription, 2–4 h are usually sufficient.

5. An S1 time-course should be performed as shown in Fig. 3 for in vitro synthesized RNA. This will establish the conditions for which all single-stranded nucleic acids, including the free probe, are completely digested. The optimal cutting time for total cytoplasmic and poly A$^+$ RNA is usually longer (30 min) than for in vitro synthesized RNA (15 min).

References

1. Berk, A. J. and Sharp, P. A. (1977) Sizing and mapping of early adenovirus mRNAs by gel electrophoresis of S1 endonuclease digested hybrids. *Cell* **12,** 721–732.
2. Favaloro, J., Treisman, R., and Kamen, R. (1980) Transcription maps of polyoma virus-specific RNA: analysis by two-dimensional nuclease S1 gel mapping. *Meth. Enzymol.* **65,** 718–749.
3. Weaver, R. F. and Weissmann, C. (1979) Mapping of RNA by a modification of Berk-Sharp procedure: the 5' termini of 15 S β-globin mRNA precursor and mature 10 S β–globin mRNA have identical map coordinates. *Nucleic Acids Res.* **7,** 1175–1193.
4. Lopata, M. A., Sollner-Webb, B., and Cleveland, D. W. (1985) Surprising S1-resistant trimolecular hybrids: potential complication in interpretation of S1 mapping analysis. *Mol. Cell. Biol.* **5,** 2842–2846.
5. Viville, S., Jongeneel, V., Koch, W., Mantovani, R., Benoist, C., and Mathis, D. (1991) The Eα promoter: a linker-scanning analysis. *J. Immunol.* **146,** 3211–3217.
6. Dean, M. (1987) Determining the hybridization temperature for S1 nuclease mapping. *Nucleic Acids Res.* **15,** 6754.

28

Measurements of Rate of Transcription in Isolated Nuclei by Nuclear "Run-Off" Assay

Rai Ajit K. Srivastava and Gustav Schonfeld

1. Introduction

Unlike gene expression in procaryotic cells, which is primarily under transcriptional control, gene expression in eucaryotic cells is subject to both transcriptional and post-transcriptional controls. Since transcription and translation in eukaryotic cells are separated topographically, the regulation of mRNA metabolism can occur at multiple sites, within nuclei and in the cytoplasm. Nevertheless, transcription remains a critical locus of control of eukaryotic gene expression. Transcriptional regulation affects cellular mRNA abundance by affecting rates of transcription. Another important control mechanism that can affect mRNA abundance is the rate of mRNA decay. Thus, half-life of mRNA represents a balance between the rates of transcription and intracellular degradation, e.g., an increase in the abundance of mRNA could result from decreased mRNA degradation, increased mRNA synthesis, or both.

Tissue specific expression and the level of expression of certain genes are routinely determined by measuring the abundance of the corresponding mRNA *(1–3)* and, frequently, mRNA levels are interpreted as reflecting rates of transcription. This is a safe interpretation in some cases *(4–8)*, but not in all cases *(9–15)*. It is, therefore, important to measure the rate of transcription directly, in order to understand the contribution of transcription rates to "setting" the cellular level of any specific mRNA.

Relative rates of transcription are measured by nuclear "run-off" assays using isolated cell nuclei. The assay quantifies the elongations in vitro of nascent mRNA chains already initiated in vivo. Nuclei are isolated from homogenized tissues or cultured cells, and incubated in the presence of four ribonucleotides, one of them radiolabeled. The reaction is stopped by the

From: *Methods in Molecular Biology, Vol. 86: RNA Isolation and Characterization Protocols*
Edited by: R. Rapley and D. L. Manning © Humana Press Inc., Totowa, NJ

addition of RNase-free DNase I, and total nuclear RNA that contains newly synthesized labeled RNA is extracted. The extracted RNA is hybridized to a denatured, immobilized cDNA corresponding to the mRNA being measured. After hybridization, the membrane is treated with RNase A and washed to remove the nonspecifically bound RNA. Bound mRNA are hydrolyzed from the filter and counted in a liquid scintillation counter. Nuclear transcription rates of given mRNAs are frequently compared to transcription rates of an "internal standard" such as, β-actin mRNA.

2. Materials

All glassware used for nuclear run-off assay is treated with 0.05% DEPC (diethyl pyrocarbonate) overnight and autoclaved for 1 h. The reagents used are also treated in a similar way and either autoclaved or filter sterilized. The reagents required are spermine, spermidine, creatine phosphate, β-mercapoethanol, HEPES (N-[2-hydroxyethyl]piperazine-N'-[ethanesulfonic acid], recombinant plasmid containing appropriate cDNA fragment, EGTA [ethylene glycol-bis β-amino ethyl] ether), dithiotheritol, nylon cloth (mesh with 40 μM), phenyl methyl sulfonyl fluoride, pancreatic ribonuclease A, heparin sulfate, ribonucleotides (GTP, ATP, CTP), ribonuclease inhibitor (RNasin or human placental ribonuclease inhibitor), α-^{32}P[UTP] (600 μCi/mM, ICN, Biomedicals), DNase I (RNase-free), guanidinium thiocyanate, sodium citrate, sacrcosyl, salt saturated phenol, chloroform/isoamylalcohol (49:1), α-amanatin, isopropanol, deionized formamide, PIPES (piperazine-N,N'-bis[2-ethane sulfonic acid]), sodium dodecyl sulfate, ficoll, polyvinylpyrrolidone, bovine serum albumin, and slmon sperm DNA.

For the isolation of nuclei following reagents are kept ready:

1. Buffer A: 60 mM KCl; 15 mM NaCl; 0.15 mM spermine; 0.5 mM Spermidine; 14 mM β-mercaptoethanol; 0.5 mM EGTA; 2 mM EDTA; and 15 mM HEPES (pH 7.5).
2. Buffer B: Same as buffer A but with 0.1 mM of EGTA and EDTA. Note: These buffers are made in DEPC-treated water and sterilized by filtration through a sterile millipore filter.)
3. Sucrose solutions: Solution A: 0.3M sucrose in buffer A; solution B: 1M sucrose in buffer B; and solution C: 1.5M sucrose solution in buffer B.
4. Nuclei storage buffer: 20 mM Tris-HCl, pH 7.9; 75 mM NaCl; 0.5 mM EDTA; 0.85 mM DTT; 0.125 mM PMSF; and 50% glycerol.
5. Wash buffer: 20 mM Tris–HCl, pH 7.5; 15 mM NaCl; and 1.1 mM sucrose.
6. Hypotonic buffer: 20 mM Tris-HCl, pH 8.0; 4 mM MgCl$_2$; 6 mM CaCl$_2$; and 0.5 mM DTT.

The above buffers and solutions are made and stored in the refrigerator (0–4°C).

7. Lysis buffer (prepare fresh from stock solutions): 0.6M sucrose; 0.2% Nonidet P-40; and 0.5 mM DTT.

8. Elongation buffer (2X): Make fresh each time from the stock solutions: 200 mM Tris-HCl, pH 7.9; 100 mM NaCl; 0.8 mM EDTA; 0.2 mM phenylmethyl sulphonyl chloride; 2.4 mM DTT; 2 mg/mL heparin sulphate; 4 mM MnCl$_2$; 8 mM MgCl$_2$; and 20 mM creatine phosphate.

9. Nucleotide mix: Prepare 100 mM solution of GTP, ATP, and CTP and mix in 1:1:1 ratio. Store frozen in aliquots of 500 μL at −20°C.

10. SET buffer (prepare fresh): 5% sodium dodecyl sulfate; 50 mM EDTA; and 100 mM Tris-HCl, pH 7.4.

11. Extraction buffer (store at room temperature): 4M guanidinium thiocyanate; 25 mM sodium citrate, pH 7.0; 0.5% Sarkosyl; and 0.1M β-mercaptoethanol.

12. Hybridization buffer (made fresh before use): 20 mM PIPES (piperazine--N,N'-bis[2-ethanesulfonic acid]), pH 6.7; 50 % formamide (deionized with mixed bed resins); 2 mM EDTA; 0.8M NaCl; 0.2% SDS; 0.02% ficoll; 0.02% polyvinylpyrrolidone; 0.02% bovine serum albumin; and 500 μg/mL denatured salmon sperm DNA.

3. Methods

3.1. Isolation of Nuclei

3.1.1. Preparation of Nuclei from Tissues

1. Homogenize preweighed tissue in 10 vol of ice-cold sucrose solution A (10 mL solution A/1 g tissue). Perform all the steps of nuclei preparation at 0–4°C.

2. Filter the homogenate through either 40-μM nylon cloth or through 4 layers of cheesecloth, and layer over a 10-mL cushion of solution B. Spin for 20 min at 2500g and 4°C in a swinging bucket rotor of Beckman (Fullerton, CA) Centrifuge.

3. Resuspend pelleted crude nuclei in solution B, using 2–3 mL for up to 1 g of tissue sample, layer over 5 mL of solution C, and centrifuge at 45,000g in Beckman SW50 rotor at 4°C for 60 min.

4. Resuspend pellet containing clean nuclei in nuclear storage buffer. This gives nuclei at a concentration of approx 10^6/μL. Isolated clean nuclei are either used immediately after preparation or may be stored frozen in aliquots of about 1–5 × 10^7 at −70°C without loss in activity for up to 4 wk.

3.1.2. Preparation of Nuclei from Cultured Cells

1. Wash a confluent culture containing 1–2 × 10^8 cells with ice-cold wash solution and collect by centrifugation at 300g.

2. Resuspend the cells in 2.5 mL ice-cold hypotonic buffer and allow to sit on ice for 5 min. Add 2.5 mL of lysis buffer.

3. Break the cells further with the tight-fitting pestle of a homogenizer. Usually 6–10 strokes are enough to break the cells.

4. Pellet nuclei by centrifugation at 1500g for 10 min and resuspend in 2 mL of sucrose solution A.

5. Layer the crude nuclear suspension over 2.5 mL of sucrose solution C and centrifuge for 1 h at 45,000g. Resuspend the nuclear pellet in a nuclear storage

buffer and either use immediately for in vitro nuclear run-off assay or store at –70°C for up to 4 wk.

3.2. Elongation of Nascent mRNA Chains

1. To a total volume of 200 μL add 100 μL of 2X elongation buffer, 6 μL of nucleotide mix, 40 U (1 μL) RNase inhibitor, 10^7 nuclei (suspended in nuclei storage buffer), and 100 μCi ^{32}P[UTP].
2. Allow the transcription reaction to proceed at 26°C for 20 min and then stop by the addition of 100 U DNase I (RNase-free). Incubate for an additional 5 min. To determine exclusively the RNA polymerase II dependent transcription, transcription is also performed in presence of α-amanitin (2 μg/mL) and the counts obtained are subtracted from the total counts obtained.
3. Treat the samples with 2 μL proteinase K (10 mg/ mL) and 20 μL SET buffer for 30 min at 37°C. Add 400 μL extraction buffer and 80 μL of sodium acetate (2.0M, pH 4.0) and vortex the contents for 10 s. Add 700 μL of salt saturated phenol and 150 μL of chloroform:isoamylalcohol, vortex for 10 s and allow to sit on ice for 15 min.
4. Centrifuge at 4°C and 12,000g for 15 min. Transfer the aqueous (top) phase to a clean tube and add 20 μg yeast tRNA and an equal volume of cold isopropanol. Mix tube contents and incubate at –20°C for 20 min. Centrifuge at 12,000g and 4°C for 15 min and wash the pellet with 70% ethanol. Lyophilize and redissolve in 100 μL of hybridization buffer (*see* **Note 1**).

3.3. Hybridization

The amount of specific mRNA synthesized in the nuclear run-off assay is determined by hybridizing the total synthesized RNA with the specific cDNA probe (*see* **Notes 2** and **3**). This is then compared to transcription of an internal control, e.g., of β-actin mRNA.

1. The double-stranded recombinant plasmid containing the appropriate cDNA fragment (or β-actin cDNA) is denatured with 0.2M NaOH/2 mM EDTA by incubation for 15 min at 37°C. It is then neutralized by 1M HEPES (N-[2-Hydroxyethyl]piperazine-N'-[2-ethane sulfonic acid]) buffer (pH 6.5).
2. Five μg of denatured plasmid DNA is applied to nitrocellulose paper using dot blot apparatus and baked for 2 h at 80°C in vacuum. The portions of the filter in which plasmid DNA are spotted are cut out with the help of a sterile cork borer or sharp blade.
3. Prehybridize the membrane for 2 h in 400 μL hybridization buffer. Remove buffer from the tube and replace with total RNA (10^5–10^6 cpm) dissolved in 250 μL of hybridization buffer. Cover hybridization mixture plus membrane with 50–100 μL of mineral oil and hybridize for 50 h at 42°C.
4. After the hybridization, wash the membranes twice with 500 μL of 2X SSC/0. 1% SDS for 30 min at room temperature. In order to remove nonspecific binding, treat the membrane with 400 μL of RNase A solution (10 μg/mL in 2X

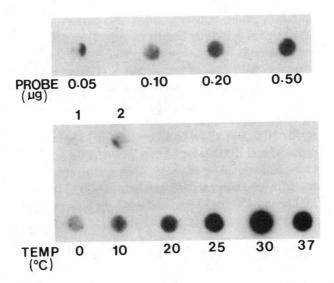

Fig. 1. Nuclear run-off assay, using 10^7 mouse liver nuclei per reaction, was performed as described in the text and probed for the rate of low-density lipoprotein receptor mRNA transcription using 1.3 kb rat liver LDL-receptor cDNA. The top panel shows an increasing hybridization signal with increasing amounts of probe immobilized onto nitrocellulose membrane. Lower panel shows the effect of temperature on the LDL-receptor mRNA transcription. Lane 1, 5 μg plasmid DNA that do not contain LDL-receptor cDNA; Lane 2, transcription performed in the presence of 2 μg/mL α-amanitin.

SSC) at 37°C for 30 min. This will remove unhybridized RNA. Wash with 1 mL of 2X SSC twice at room temperature and expose to X-ray film (*see* **Figs. 1** and **2**).

5. To elute the hybridized RNA, incubate the membranes with 200 μL of 0.3*M* NaOH for 15 min at 65°C followed by the addition of 50 μL of glacial acetic acid and 4 mL of scintillation cocktail. Count ^{32}P radioactivity in a liquid scintillation counter. This value provides the relative rates of transcription of a specific mRNA (*see* **Note 4**).

4. Notes

1. The integrity of newly synthesized RNA and the extent of incorporation of label in the nascent RNA chains may be checked prior to proceeding for hybridization:

   a. To determine the extent of label incorporation, an aliquot of synthesized total nuclear RNA is diluted 10–20-fold depending on the label incorporation as judged by Geiger counter. An aliquot (usually 2 concentrations; 5 and 10 μL) of the diluted sample is counted in a liquid scintillation counter.

   b. To determine the quality of the synthesized RNA, an aliquot of newly synthesized RNA (50,000–100,000 cpm) is resolved in 6% denaturing polyacrylamide gel containing 7*M* urea. After the electrophoresis, the gel is dried and exposed to X-ray film. If the transcribed RNAs are intact, one can

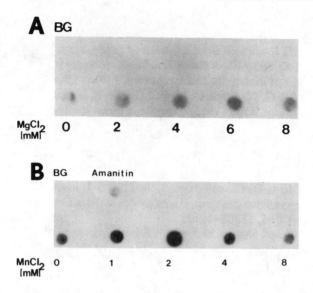

Fig. 2. LDL-receptor mRNA transcription on isolated mouse liver nuclei in the presence of different concentrations of $MgCl_2$ or $MnCl_2$. Each transcription assay was performed with 10^7 nuclei at 28°C for 20 min. The concentration of divalent cations are indicated in the figure.

    see several distinct RNA bands all along the autoradiogram. If the RNAs are degraded, smaller fragments of RNA appear and there is smudging of the bands.

2. To ensure specificity of the probe, linear concentrations (0.1–1 µg) of recombinant plasmid are immobilized on one set of filters, and identical amounts of a plasmid that does not contain cDNA insert are bound to another set of nitrocellulose membranes. Hybridization is performed with the same amounts of $^{32}P[RNA]$ using both sets of filters. If the cDNA probe is specific, a linear increase in the hybridization signal is obtained with the recombinant plasmid but not with the nonrecombinant plasmid.

3. In another experiment, increasing amounts of $^{32}P[RNA]$ (10^5–10^6 cpm) are hybridized with constant amounts (5 µg) of the recombinant and the nonrecombinant plasmids. Here again one gets increasing hybridization signals with the recombinant plasmid but not with the nonrecombinant plasmid.

4. For background counts hybridization is carried with the nonrecombinant plasmid and the counts obtained are subtracted from the counts obtained by hybridization with the recombinant plasmid.

References

1. Sorci-Thomas, M., Wilson, M. D., Johnson, F. L., Williams, D. L., and Rudel, L. L. (1989) Studies on the expression of genes encoding apolipoproteins B-100 and B-48 and the low density lipoprotein receptor in non-human primates. Comparison of dietary fat and cholesterol. *J. Biol. Chem.* **264**, 9039–9045.

2. Srivastava, R. A. K., Ito, H., Hess, M., Srivastava, N., and Schonfeld, G. (1995) Regulation of low density lipoporotein gene expression in HepG2 and Caco2 cells by palmitate, oleate, and 25-hydroxycholesterol. *J. Lipid Res.* **36,** 1434–1446.

3. Srivastava, R. A. K., Baumann, D., and Schonfeld, G. (1993) In vivo regulation of low density lipoprotein receptore by estrogen differs at the posttranscriptional level in rat and mouse. *Eur. J. Biochem.* **216,** 527–538.

4. Srivastava, R. A. K. (1996) Regulation of apolipoprotein E by dietary lipids occurs by transcriptional and posttranscriptional mechanisms. *Mol. Cell Biochem.* **155,** 153–162.

5. Brock, M. L. and Shapiro, D. J. (1983) Estrogen regulates the absolute rate of transcription of the *Xenopus laevis* vitellogenin genes. *J. Biol. Chem.* **258,** 5449–5455.

6. McKnight, G. S. and Palmiter, R. D. (1979) Transcriptional regulation of the ovalbumin and conalbumin genes by steroid hormones in chick oviduct. *J. Biol. Chem.* **254,** 9050–9058.

7. Chazenbalk, G. D., Wadsworth, H. L., and Rapoport, B. (1990) Transcriptional regulation of ferritin H messenger RNA levels in FRTL5 rat thyroid cells by thyrotropin. *J. Biol. Chem.* **265,** 666–670.

8. Chinsky, J. M., Maa, M. C., Ramamurthy, V., and Kellems, R. E. (1989) Adenosine deaminase gene expression. Tissue-dependent regulation of transcriptional elongation. *J. Biol. Chem.* **264,** 14,561–14,565.

9. Saini, K., Thomas, P., and Bhandari, B. (1990) Hormonal regulation of stability of glutamine synthetase mRNA in cultured 3T3-L1 adipocytes. *Biochem. J.* **267,** 241–244.

10. Brock, M. L. and Shapiro, D. J. (1983) Estrogen stabilizes vitellogenin mRNA against cytoplasmic degradation. *Cell* **34,** 207–214.

11. Antrast, J., Lasnier, F., and Pairault, J. (1991) Adipsin gene expression in 3T3-F442A adipocytes is post-transcriptionally down-regulated by retinoic acid. *J. Biol. Chem.* **266,**1157–1161.

12. Jefferson, D. M., Clayton, D. F., Darnell, J. E., Jr., and Reid, L. M. (1986) Posttranscriptional modulation of gene expression in cultured rat hepatocytes. *Mol. Cell Biol.* **4,** 1929–1934.

13. Hod, Y. and Hanson, R. W. (1988) Cyclic AMP stabilizes the mRNA for phosphoenol pyruvate carboxykinase (GTP) against degradation. *J. Biol. Chem.* **263,** 7747–7752.

14. Srivastava, R. A. K., Kitchens, R. T., and Schonfeld, G. (1994) Regulation of the apolipoprotein AIV gene expression by estrogen differs in rat and mouse. *Eur. J. Biochem.* **222,** 507–514.

15. Srivastava, R. A. K., Jiao, S., Tang, J., Pfleger, B., Kitchens, R. T., and Schonfeld, G. (1991) In vivo regulation of low density lipoprotein receptor and apolipoprotein B gene expression by dietary fatty acids and dietary cholesterol in inbred strains of mice. *Biochem. Biophys. Acta* **1086,** 29–43.

29

Transcription In Vitro Using Bacteriophage RNA Polymerases

Elaine T. Schenborn

1. Introduction

Synthesis of specific RNA sequences in vitro is simplified because of the availability of bacteriophage RNA polymerases and specially designed DNA vectors. RNA polymerases encoded by SP6, T7, or T3 bacteriophage genomes recognize particular phage promoter sequences of their respective viral genes with a high degree of specificity *(1–3)*. These RNA polymerases also transcribe DNA templates containing their cognate promoters under defined conditions in vitro *(4,5)*. Standard reaction conditions for transcription in vitro can be adjusted for synthesis of large amounts of RNA or for smaller amounts of labeled RNA probes (*see* Chapters 30–32).

Larger-scale in vitro synthesis produces RNA that mimics biologically active RNA in many applications. The following examples represent some of the different uses for RNA synthesized in vitro. RNA transcripts are particularly well suited for the study of RNA virus gene regulation. For example, the in vitro transcribed RNA genomes of poliovirus *(6)* and cowpea mosaic virus *(7)* produce infectious particles in transfected cells. For other types of studies, messenger RNA-like transcripts are used as substrates to study RNA processing activities, such as splicing *(8)* and 3'-end maturation *(9,10)*. RNA transcripts synthesized in vitro are also widely used as templates for protein synthesis in cell-free extracts designed for in vitro translation *(11)*. Transfer RNA-like transcripts have been used as substrates to study RNase P cleavage specificities *(12)*, and other mechanisms of RNA cleavage have been investigated using RNA substrates and ribozymes synthesized in vitro *(13)*. Gene regulation studies using antisense RNA also have taken advantage of the ease of in vitro RNA synthesis. In vitro translation of a targeted message has been shown to be

From: *Methods in Molecular Biology, Vol. 86: RNA Isolation and Characterization Protocols*
Edited by: R. Rapley and D. L. Manning © Humana Press Inc., Totowa, NJ

inhibited in the presence of antisense RNA in vitro *(14)*, and in vivo translation has been blocked in *Xenopus* oocytes by antisense RNA *(15)*. The ability to synthesize discrete RNA templates in vitro also facilitates studies of RNA and protein interactions *(16,17)*.

The generation of radioactively labeled RNA hybridization probes is a widely used application for RNA synthesized in vitro. RNA probes are synthesized predominantly by incorporation of a radiolabeled ribonucleotide, ^{32}P-, ^{3}H-, or ^{35}S-rNTP, into the transcript. Nonisotopic probes can be synthesized by incorporation of biotinylated *(18)* or digoxigenin *(19)* modified bases. For Northern blots, single-stranded RNA probes are generally more sensitive than the corresponding DNA probe because of the higher thermal stability of RNA:RNA hybrids compared to RNA:DNA hybrids and the absence of self-complementary sequences in the probe preparation *(4)*.

RNA probes also are more sensitive than DNA probes for the detection of DNA sequences transferred to membranes from Southern blots, plaque lifts, and colony lifts *(20)*. The lower background and increased signal sensitivity of RNA probes are possible because of higher stability of RNA:DNA hybrids compared to DNA:DNA hybrids. This increased stability allows more stringent conditions to be used for the hybridization and washing procedures *(21)*. Another advantage of RNA probes is that RNase A can be added after the hybridization reaction to eliminate nonspecific binding of the probe to the membrane. High sensitivity also has been achieved with RNA probes used for *in situ* hybridization *(22)* and localization of genes in chromosome spreads *(23)*. RNase mapping is another application that takes advantage of the superior properties of RNA probes for hybridization to complementary sequences. In this application, a radiolabeled RNA probe is hybridized in solution to cellular RNA, then the nonhybridized, single-stranded regions of the probe are later digested with RNase A and RNase T1, and the protected, hybridized regions are identified by gel analysis. This type of mapping is used to quantitate low-abundance species of RNA, and to map exons, transcription start sites, and point mutations *(4,24)*.

The DNA templates used for in vitro transcription contain the cloned sequence of interest immediately "downstream" of an SP6, T7, or T3 promoter sequence. Plasmid vectors are commercially available with the phage promoter sequence adjacent to a cloning region. One example is the pGEM® series of vectors (Promega, Madison, WI) designed with multiple cloning sites flanked by opposed SP6 and T7 promoters, allowing the synthesis of either sense or antisense RNA from a single recombinant plasmid. Discrete RNAs, corresponding to the cloned sequence of interest, are synthesized as "run-off" transcripts from a linear DNA template. To prepare the linear template, the recombinant plasmid DNA is cut with a restriction enzyme cleaving within, or shortly downstream of, the cloned insert. The linear DNA is then added to the reaction mixture for in vitro synthesis of RNA (*see* **Fig. 1**).

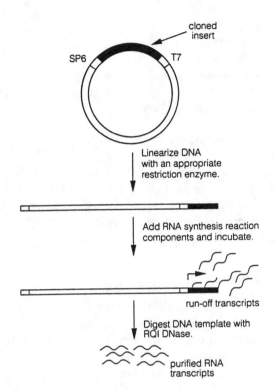

Fig. 1. Synthesis of RNA by transcription in vitro from a linear DNA template.

2. Materials

1. Transcription buffer (5X): 200 mM Tris-HCl, pH 7.5, 30 mM MgCl$_2$, 10 mM spermidine, and 50 mM NaCl. Store at –20°C.
2. ATP, GTP, CTP, UTP: 10 mM stocks prepared in sterile, nuclease-free water and adjusted to pH 7.0. Store at –20°C.
3. 100 mM DTT: Store at –20°C.
4. RNasin® Ribonuclease Inhibitor: (Promega) Store at –20°C.
5. Nuclease-free water: Prepare by adding 0.1% diethyl pyrocarbonate (DEPC) to the water. Autoclave to remove the DEPC. **Caution:** DEPC is a suspected carcinogen.
6. TE buffer: 10 mM Tris-HCl, pH 8.0, and 1 mM EDTA. Prepare with stock solutions that are nuclease-free.
7. TE-saturated phenol/chloroform: Mix equal parts of TE buffer and phenol, and allow phases to separate. Mix 1 part of the lower, phenol phase with 1 part of chloroform:isoamyl alcohol (24:1).
8. Chloroform:isoamyl alcohol (24:1): Mix 24 parts of chloroform with 1 part isoamyl alcohol.

9. Ammonium acetate: 7.5 and 2.5*M*.
10. 3*M* sodium acetate, pH 5.2.
11. Ethanol: Absolute (100%) and 70%.
12. Enzymes: SP6, T3, or T7 RNA polymerase at 15–20 U/µL.
13. RNase-free DNase: RQ1 (Promega).
14. Restriction enzyme and appropriate buffer to linearize plasmid DNA template.
15. DE–81 filters: 2.4 cm diameter (Whatman).
16. 0.5*M* Na_2HPO_4, pH 7.0.
17. $m^7G(5')ppp(5')G$: 5 m*M* (New England BioLabs).

Microcentrifuge tubes, pipet tips, glassware: To provide a nuclease-free environment, use sterile, disposable microcentrifuge tubes and pipet tips whenever possible for the preparation and storage of reagents. Larger volumes of reagents can be stored in bottles that have been baked at 250°C for four or more hours to inactivate RNases.

3. Methods

Throughout these procedures, precautions should be taken to protect against ribonuclease contamination. These precautions include the use of sterile, nuclease-free reagents and materials, and the use of disposable gloves to prevent accidental contamination of samples with ribonucleases present on the skin.

Three steps are required for synthesis of RNA in vitro:

1. Preparation of the DNA template.
2. Transcription reaction.
3. Enrichment of the RNA product.

3.1. Preparation of the DNA Template

The sequence of interest is cloned by established methods into an appropriate vector, downstream of a promoter sequence for SP6, T7, or T3 RNA polymerase. The recombinant plasmid DNA is purified, and either added directly to the in vitro transcription reaction or linearized prior to the run-off transcription reaction. Transcription of supercoiled plasmid DNA results in the synthesis of high-mol-wt RNA, which contains vector sequences. Discrete RNA sequences of interest, without vector sequence, are generated by run-off transcription from linear templates prepared in the following manner:

1. Determine the restriction site downstream of, or within, the cloned insert, which will generate the desired run-off transcript. Whenever possible, select a restriction enzyme that produces 5' overhanging or blunt ends. If an enzyme that generates a 3' overhang is selected, *see* **Note 1**. Set up the restriction digest according to the enzyme supplier's directions.

2. Check for completeness of digestion by agarose gel electrophoresis. During this analysis, keep the DNA sample on ice. If digestion is complete, proceed with **step 3**. Otherwise, add additional restriction enzyme to the DNA, incubate an additional 30 min, and repeat the agarose gel analysis. Ensure that restriction enzyme does not exceed 10% of final volume. Restriction enzymes are supplied in glycerol which at high concentration can inhibit its activity.

3. Extract the DNA by adding an equal volume of TE-saturated phenol/chloroform, vortex for 1 min, and centrifuge at 12,000g for 2 min. Transfer the upper phase to a fresh tube, and add 1 vol of chloroform:isoamyl alcohol (24:1). Vortex for 1 min, and centrifuge at 12,000g for 2 min.

4. Precipitate the DNA by transferring the upper, aqueous phase to a fresh tube, and adding 0.1 vol of 3M sodium acetate, pH 5.2, and 2 vol of absolute ethanol. Cool 30 min at −70°C, and centrifuge at 12,000g for 5 min.

5. Carefully pour off the supernatant, wash the pellet briefly with 1 mL of 70% ethanol, spin at 12,000g for 2 min, and remove the supernatant. Dry briefly in a vacuum desiccator. Resuspend the pellet in nuclease-free water or TE buffer to a final DNA concentration of approx 1 mg/mL.

3.2. Synthesis of Radiolabeled RNA Probes (see Notes 2–5)

RNA probes at a specific activity of 6–9 × 10^8 cpm/µg can be generated by transcribing DNA in the presence of a limiting concentration (12–24 µM) of one radiolabeled ribonucleotide and saturating concentrations (0.5 mM) of the other three rNTPs (*see* **Notes 2** and **3**). The following example uses 50 µCi of α-[^{32}P]CTP at a specific activity of 400 Ci/mmol/20 µL reaction, providing a final concentration of 6 µM of α-[^{32}P]CTP. An additional 12 µM of unlabeled CTP is added to bring the total concentration to 18 µM CTP. Expect approx 1 mol of RNA/mol of DNA template to be synthesized under these conditions.

1. To a sterile microcentrifuge tube, add the following components at room temperature in the order listed. This order of addition prevents precipitation of the DNA by spermidine: 4 µL of 5X transcription buffer, 2 µL of 100 mM DTT, 20 U RNasin® Ribonuclease Inhibitor, 4 µL of ATP, GTP, and UTP (2.5 mM each; prepare by mixing 1 vol of each individual 10 mM stock of ATP, GTP, and UTP, and 1 vol of water), 2.4 µL of 100 µM CTP (dilute 10 mM stock 1:100 with water), 1 µL of DNA template (up to 2 µg; 1–2 mg/mL in nuclease-free water or TE), 5 µL of a-[^{32}P]CTP (400 Ci/mmol; 10 mCi/mL). Bring to a final vol of 19 µL with nuclease-free water.

2. Initiate the reaction by adding 1 µL of SP6, T7, or T3 RNA polymerase (at 15–20 U/µL).

3. Incubate for 60 min at 37–40°C.

4. Remove 1 µL from the reaction at this point to determine the percent incorporation and specific activity of the probe. The remainder of the sample can be digested by RQ1 RNase-free DNase (**Subheading 3.6.**).

3.3. Determination of Percent Incorporation and Probe Specific Activity

1. Remove 1 μL of the labeled probe, and dilute into 19 μL of nuclease-free water. Spot 3 μL of this 1:20 dilution onto 4 DE81 filters. Dry the filters at room temperature or under a heat lamp.
2. Place two filters directly into separate scintillation vials, add scintillation fluid, and count. Calculate the average cpm per filter, and determine the total cpm per microliter of original reaction as follows:

$$\text{Total cpm/μL of original reaction} = \text{average cpm per filter} \times (\text{20-fold dilution/3 μL}) \tag{1}$$

3. Wash the unincorporated nucleotides from the remaining two filters by placing the filters in a small beaker containing 50–100 mL of 0.5M Na$_2$HPO$_4$ (pH 7.0). Swirl the filters occasionally for 5 min, then decant, and replace with fresh buffer. Repeat the wash procedure two more times. Dip the filters briefly into 70% ethanol, and dry at room temperature or under a heat lamp.
4. Place each filter into a scintillation vial, add scintillation fluid, and count. Calculate the amount of labeled nucleotide incorporated into RNA (incorporated cpm) per microliter of original reaction as follows:

$$\text{Incorporated cpm/μL of original reaction} = \text{average cpm per filter} \times (\text{20-fold dilution/3 μL}) \tag{2}$$

This value will also be used in estimating the probe specific activity in **step 6**.
5. Calculate the percent incorporation from the values determined above in **steps 2** and **4**.

$$\text{\% Incorporation} = (\text{incorporated cpm/total cpm}) \times 100 \tag{3}$$

The percentage of incorporation under the conditions described generally ranges from 70 to nearly 100%. A low incorporation of radiolabeled nucleotide (for example, below 50%) reflects a low yield of RNA product (*see* **Note 5**).
6. Calculate the specific activity of the probe as cpm/μg RNA synthesized. To do this, first calculate the total incorporated cpm in the reaction:

$$\text{Total incorporated cpm} = (\text{incorporated cpm/μL of reaction}) \times 20\ \text{μL reaction vol} \tag{4}$$

Next we need to calculate the total nmoles of nucleotide in the reaction to determine how many micrograms of RNA were synthesized; 50 μCi of α-[^{32}P]CTP at 400 μCi/nmol corresponds to 0.12 nmol of ^{32}P-CTP/reaction. Adding in the 12 μM of unlabeled CTP (0.24 nmol) gives a total of 0.36 nmol of CTP. If a maximum 100% incorporation was achieved and CTP represents one-fourth of all the nucleotides in the probe, then the total amount of nucleotides incorporated into the probe would be (0.36 nmol × 4) or 1.44 nmol. Assuming an average FW/nucleotide of 330, the amount of RNA synthesized in this example would be 1.44 nmol × (330 ng/nmol) = 475 ng of RNA synthesized. If the percentage of incorporation calculated from **step 5** was 80%, for example, then the actual amount of RNA synthesized in the reaction would be 475 ng × 0.80 = 380 ng RNA.

$$SA = \text{total incorporated cpm/µg RNA} \tag{5}$$

In this example, the total incorporated CPM would be divided by 0.380 µg RNA.

3.4. Synthesis of Large Quantities of RNA (see Notes 2–6)

Using the following reaction conditions in which all four rNTPs are at a saturating concentration, yields of 5–10 µg of RNA/µg of DNA template can be obtained (*see* **Note 6**). This represents up to 20 mol of RNA/mol of DNA template. Incubation with additional polymerase after the initial 60-min reaction can increase the yield of RNA up to twofold. The following reaction can be scaled up or down as desired.

1. To a sterile microcentrifuge tube, add the following components at room temperature in the order listed. This order of addition prevents precipitation of the DNA by spermidine: 20 µL of 5X transcription buffer, 10 µL of 100 m*M* DTT, 100 U RNasin Ribonuclease Inhibitor, 20 µL of ATP, GTP, UTP, and CTP (2.5 m*M* each; prepare by mixing 1 vol of each individual 10 m*M* stock of ATP, GTP, UTP, and CTP), 2–5 µL of DNA template (5–10 µg total; 1–2 mg/mL in nuclease-free water or TE). Add nuclease-free water to a final vol of 98 µL.
2. Initiate the reaction by adding 2 µL of SP6, T7, or T3 RNA polymerase (at 15–20 U/µL).
3. Incubate for 60 min at 37–40°C.
4. Add an additional 2 µL of SP6, T7, or T3 RNA polymerase. Incubate for 60 min at 37–40°C.

The DNA template can now be digested by RQ1 RNase-free DNase (**Subheading 3.6.**).

3.5. Synthesis of 5′ Capped Transcripts

Some RNA transcripts require a m⁷G(5′)ppp(5′)G cap at the 5′ end for higher translation efficiency, either in cell-free extracts or in *Xenopus* oocytes *(25)*. Methylated capped transcripts also have been reported to function more efficiently for in vitro splicing reactions *(8)* and are more resistant to ribonucleases in nuclear extracts. The following reaction can be scaled up or down as desired.

1. To a sterile microcentrifuge tube, add the following components at room temperature in the order listed. This order of addition prevents precipitation of the DNA by spermidine: 4 µL of 5X transcription buffer, 2 µL of 100 m*M* DTT, 20 U RNasin Ribonuclease Inhibitor, 4 µL of ATP, UTP, and CTP (2.5 m*M* each; prepare by mixing 1 vol of each individual 10 m*M* stock of ATP, UTP, and CTP, and 1 vol of water), 2 µL of GTP (0.5 m*M*; dilute 10 m*M* stock 1:20 with water),

2 µL of the cap analog m^{7}G(5')ppp(5')G (5 mM), and 1 µL of DNA template: 1–2 µg (1–2 mg/mL in nuclease-free water or TE). Add nuclease-free water, if necessary, to a final vol of 19 µL.

2. Initiate the reaction by adding 1 µL of SP6, T7, or T3 RNA polymerase (at 15–20 U/µL).
3. Incubate for 60 min at 37–40°C.

The DNA template can now be digested by RQ1 RNase-free DNase (**Subheading 3.6**).

3.6. Digestion of the DNA Template Post-Transcription

To achieve maximal sensitivities with RNA probes, the DNA template must be eliminated after the transcription reaction. Elimination of the DNA template also may be required for the preparation of biologically active RNAs. DNase can be used to digest the DNA template, but during this enzymatic step, it is critical to maintain the integrity of the RNA. RQ1 DNase (Promega) is certified to be RNase-free and is recommended for the following protocol.

1. After the in vitro transcription reaction, add RQ1 RNase-free DNase to a concentration of 1 U/µg of template DNA.
2. Incubate for 15 min at 37°C.
3. Extract with 1 vol of TE-saturated phenol/chloroform. Vortex for 1 min, and centrifuge at 12,000g for 2 min.
4. Transfer the upper, aqueous phase to a fresh tube. Add 1 vol of chloroform:isoamyl alcohol (24:1). Vortex for 1 min and centrifuge as in **step 3**.
5. Transfer the upper, aqueous phase to a fresh tube. At this point, a small aliquot can be taken for electrophoretic analysis on a denaturing gel, and the remainder of the sample can be precipitated (**Subheading 3.7.**).

3.7. Precipitation of RNA

1. Add 0.5 vol of 7.5M ammonium acetate to the aqueous RNA sample prepared in **Subheading 3.6**. If the RNA sample was not digested with RQ1 DNase, extract the RNA after the transcription reaction with TE-saturated phenol/chloroform followed by a chloroform extraction, as described in **Subheading 3.6., steps 3–5**.
2. Add 2.5 vol of ethanol, mix, and place at –70°C for 30 min.
3. Centrifuge at 12,000g for 5 min. Carefully remove the supernatant.
4. Resuspend the RNA pellet in 100 µL of 2.5M ammonium acetate and mix.
5. Repeat the ethanol precipitation as described in **steps 2** and **3** above.
6. Dry the pellet briefly under vacuum, and resuspend in 20 µL or other suitable volume of sterile TE or nuclease-free water. Store the RNA at –70°C.

4. Notes

1. Extraneous transcripts complementary to the opposite strand and vector sequences are generated from DNA templates with 3' overhanging ends *(26)*. The ends of

Table 1
SA and Concentration of rNTPs Used for Transcription In Vitro

Nucleotide	Specific activity	µCi/reaction	Final conc.
α-[^{32}P] rNTP	400 Ci/mmol	50 µCi	6 µM
α-[^{35}S] rNTP	1300 Ci/mmol	300 µCi	12 µM
5,6[^{3}H] rNTP	40 Ci/mmol	25 µCi	31 µM

these templates can be made blunt in the following manner using the 3'–5' exonuclease activity of the Klenow fragment of DNA polymerase I. Set up the transcription reaction, but without nucleotides and RNA polymerase. Add 5 U of Klenow fragment/µg DNA, and incubate for 15 min at 22°C. Then initiate the transcription reaction by adding nucleotides and RNA polymerase, and incubate for 60 min at 37–40°C.

2. Incomplete transcripts are more likely to be generated under the conditions used for probe synthesis, in which the concentration of a radiolabeled nucleotide becomes limiting. Of the four nucleotides, rGTP yields the highest percentage of full-length transcripts when present in limiting concentrations *(4)*. However, for best results, radiolabeled rGTP should be used within 1 wk of the reference date. rATP yields the lowest percentage of full-length transcripts and lowest incorporation when present at a limiting concentration *(5)*. In some cases, the amount of full-length transcripts increases when the incubation temperature is lowered to 30°C. Another possible cause for incomplete transcripts can be the presence of a sequence within the DNA template that acts as a terminator for that particular polymerase. In this case, one can subclone the sequence of interest behind a different RNA polymerase promoter.

3. The specific activity of a probe can be increased by using more than one radiolabeled nucleotide per reaction at a limiting concentration. Also, more than 5 µL of the radionucleotide can be used per 20 µL reaction if the nucleotide is first aliquoted into the reaction tube and dried down under vacuum. **Table 1** lists the final concentration (final conc.) of radionucleotides commonly used in RNA probe synthesis, in a 20-µL reaction volume. Thiol-substituted rNTPs are incorporated less efficiently by the RNA polymerases than the corresponding ^{32}P or [3]H rNTPs *(5)*.

4. Biotinylated rNTP can be added during the transcription reaction, but the yield of RNA may be lowered. Alternatively, RNA can be modified after transcription using photoactivatable biotin *(27)*.

5. A low yield of RNA product can be caused by several conditions, including precipitation of DNA by spermidine in the transcription buffer, RNase contamination, carryover of residual contaminants or salts in the DNA preparation, or inactive RNA polymerase.

6. High yields of RNA synthesized by SP6 or T7 RNA polymerase have been reported using a transcription buffer containing 80 m*M* HEPES-KOH, pH 7.5, 2 m*M*

spermidine, 10–40 mM DTT, 3 mM each rNTP, 12–16 mM MgCl$_2$, and 1200–1800 U/mL RNA polymerase. Under these conditions, yields up to 80 μg of RNA/μg DNA were reported *(28)*.

References

1. Butler, E. T. and Chamberlin, M. J. (1982) Bacteriophage SP6-specific RNA polymerase. *J. Biol. Chem.* **257,** 5772–5778.
2. Davanloo, P., Rosenberg, A. H., Dunn, J. J., and Studier, F. W. (1984) Cloning and expression of the gene for bacteriophage T7 RNA polymerase. *Proc. Natl. Acad. Sci. USA* **81,** 2035–2039.
3. Jorgensen, E. D., Joho, K., Risman, S., Moorefield, M. B., and McAllister, W. T. (1989) Promoter recognition by bacterophage T3 and T7 RNA polymerases, in *DNA–Protein Interaction in Transcription* (Gralla, J. D., ed.), Liss, New York, pp. 79–88.
4. Melton, D. A., Krieg, P. A., Rebagliati, M. R., Maniatis, T., Zinn, K., and Green, M. R. (1984) Efficient *in vitro* synthesis of biologically active RNA and RNA hybridization probes from plasmids containing a bacteriophage SP6 promoter. *Nucleic Acids Res.* **12,** 7035–7056.
5. Krieg, P. A. and Melton, D. A. (1987) *In vitro* RNA synthesis with SP6 RNA polymerase. *Methods Enzymol.* **155,** 397–415.
6. Kaplan, G., Lubinski, J., Dasgupta, A., and Racaniello, V. R. (1985) *In vitro* synthesis of infectious poliovirus RNA. *Proc. Natl. Acad. Sci. USA* **82,** 8424–8248.
7. Eggen, R., Verver, J., Wellink, J., DeJong, A., Goldbach, R., and van Kammen, A. (1989) Improvements of the infectivity of in vitro transcripts from cloned cowpea mosaic virus cDNA: impact of terminal nucleotide sequences. *Virology* **173,** 447–455.
8. Krainer, A. R., Maniatis, T., Ruskin, B., and Green, M. R. (1984) Normal and mutant human β-globin pre-mRNAs are faithfully and efficiently spliced *in vitro*. *Cell* **36,** 993–1005.
9. Krieg, P. A. and Melton, D. A. (1984) Formation of the 3' end of histone mRNA by post-transcriptional processing. *Nature* **308,** 203–206.
10. Georgiev, O., Mous, J., and Birnstiel, M. (1984) Processing and nucleo-cytoplasmic transport of histone gene transcripts. *Nucleic Acids Res.* **12,** 8539–8551.
11. Krieg, P. A. and Melton, D. A. (1984) Functional messenger RNAs are produced by SP6 *in vitro* transcription of cloned cDNAs. *Nucleic Acids Res.* **12,** 7057–7070.
12. Burgin, A. B. and Pace, N. R. (1990) Mapping the active site of ribonuclease P RNA using a substrate containing a photoaffinity agent. *EMBO J.* **9,** 4111–4118.
13. Heus, H. A., Uhlenbeck, O. C., and Pardi, A. (1990) Sequence-dependent structural variations of hammerhead RNA enzymes. *Nucleic Acids Res.* **18,** 1103–1108.
14. Nicole, L. M. and Tanguay, R. M. (1987) On the specificity of antisense RNA to arrest *in vitro* translation of mRNA coding for Drosophila hsp 23. *Biosci. Rep.* **7,** 239–246.
15. Melton, D. A. (1985) Injected antisense RNAs specifically block messenger RNA translation *in vivo*. *Proc. Natl. Acad. Sci. USA* **82,** 144–148.
16. Witherell, G. W., Wu, H.-N., and Uhlenbeck, O. C. (1990) Cooperative binding of R17 coat protein to RNA. *Biochemistry* **29,** 11,051–11,057.

17. Turek, C. and Gold, L. (1990) Systematic evolution of ligands by exponential enrichment: RNA ligands to bacteriophage T4 DNA polymerase. *Science* **249**, 505–510.
18. Langer, P. R., Waldrop, A. A., and Ward, D. C. (1982) Enzymatic synthesis of biotin-labeled polynucleotides: novel nucleic acid affinity probes. *Proc. Natl. Acad. Sci. USA* **78**, 6633–6637.
19. Aigner, S. and Pette, D. (1990) *In situ* hybridization of slow myosin heavy chain mRNA in normal and transforming rabbit muscles with the use of a nonradioactively labeled cRNA. *Histochemistry* **95**, 11–18.
20. Sambrook, J., Fritsch, E. F., and Maniatis, T. (1989) *Molecular Cloning, A Laboratory Manual*, 2nd ed., Cold Spring Harbor Laboratory, Cold Spring Harbor, NY.
21. Casey, J. and Davidson, N. (1977) Rates of formation and thermal stabilities of RNA:DNA and DNA:DNA duplexes at high concentrations of formamide. *Nucleic Acids Res.* **4**, 1539–1552.
22. Uhlig, H., Saeger, W., Fehr, S., and Ludecke, D. K. (1991) Detection of growth hormone, prolactin and human beta-chorionic gonadotropin messenger RNA in growth-hormone-secreting pituitary adenomas by *in situ* hybridization. *Virchows Arch. Pathol. Anat. Histopathol.* **418**, 539–546.
23. Matthaei, K. I. and Reed, K. C. (1986) Chromosome assignment in somatic hybrids by *in situ* hybridization with tritium labeled Riboprobe® RNA probes. *Promega Notes* **5**, 5–6.
24. Zinn, K., DiMaio, D., and Maniatis, T. (1983) Identification of two distinct regulatory regions adjacent to the human β-interferon gene. *Cell* **34**, 865–879.
25. Contreras, R., Cheroutre, H., Degrave, W., and Fiers, W. (1982) Simple, efficient *in vitro* synthesis of capped RNA useful for direct expression of cloned eukaryotic genes. *Nucleic Acids Res.* **10**, 6353–6362.
26. Schenborn, E. T. and Mierendorf, R. C. (1985) A novel transcription property of SP6 and T7 RNA polymerases: dependence on template structure. *Nucleic Acids Res.* **13**, 6223–6236.
27. Forster, A. C., McInnes, J. L., Skingle, D. C., and Symons, R. H. (1985) Non-radioactive hybridization probes prepared by the chemical labelling of DNA and RNA with a novel reagent, photobiotin. *Nucleic Acids Res.* **13**, 745–761.
28. Gurevich, V. V., Pokrovskaya, I. D., Obukhova, T. A., and Zozulya, S. A. (1991) Preparative *in vitro* mRNA synthesis using SP6 and T7 RNA polymerases. *Analyt. Biochem.* **195**, 207–213.

30

In Vitro Translation of Messenger RNA in a Rabbit Reticulocyte Lysate Cell-Free System

Louise Olliver and Charles D. Boyd

1. Introduction

The identification of specific messenger RNA molecules and the characterization of the proteins encoded by them has been greatly assisted by the development of in vitro translation systems. These cell-free extracts comprise the cellular components necessary for protein synthesis, i.e., ribosomes, tRNA, rRNA, amino acids, initiation, elongation and termination factors, and the energy-generating system (1). Heterologous mRNAs are faithfully and efficiently translated in extracts of HeLa cells (2), Krebs II ascites tumor cells (2), mouse L cells (2), rat and mouse liver cells (3), Chinese hamster ovary (CHO) cells (2), and rabbit reticulocyte lysates (2,4), in addition to those of rye embryo (5) and wheat germ (6). Translation in cell-free systems is simpler and more rapid (60 min vs 24 h) than the in vivo translation system using *Xenopus* oocytes.

The synthesis of mRNA translation products is detected by their incorporation of radioactively labeled amino acids, chosen specifically to be those occurring in abundance in the proteins of interest. Analysis of translation products usually involves specific immunoprecipitation (7), followed by polyacrylamide gel electrophoresis (8) and fluorography (9) (*see* **Fig. 1**).

In vitro translation systems have played important roles in the identification of mRNA species and the characterization of their products, the investigation of transcriptional and translational control, and the cotranslational processing of secreted proteins by microsomal membranes added to the translation reaction (10,11) (*see* Chapters 29, 31, and 32). This chapter describes the rabbit reticulocyte lysate system for in vitro translation of mRNA.

From: *Methods in Molecular Biology, Vol. 86: RNA Isolation and Characterization Protocols*
Edited by: R. Rapley and D. L. Manning © Humana Press Inc., Totowa, NJ

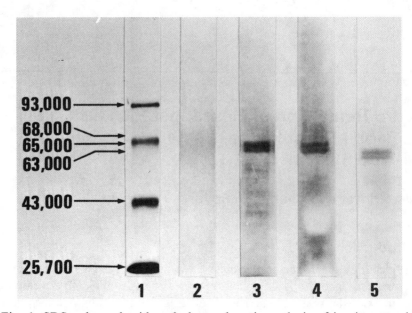

Fig. 1. SDS polyacrylamide gel electrophoretic analysis of in vitro translation products. In vitro translation products were derived from exogenous mRNA in an mRNA-dependent reticulocyte lysate cell-free system. Following electrophoresis on 8% SDS polyacrylamide gels, radioactive protein products were analyzed by flourography. Lane 1: [^{14}C]-labeled proteins of known molecular weights, i.e., phosphorylase a (93K), bovine serum albumin (68K), ovalbumin (43K), α-chymotrypsinogen (25.7K). Lanes 2–5 represent [^{3}H]-proline-labeled translation products of the following mRNAs: Lane 2: endogenous reticulocyte lysate mRNA, Lane 3: 0.3 μg of calf nuchal ligament polyadenylated RNA. Lane 4: 0.3 μg of calf nuchal ligament polyadenylated RNA, and immunoprecipitated with 5 μL of sheep antiserum raised to human tropoelastin, Lane 5: 0.3 μg of calf nuchal ligament polyadenylated RNA and cotranslationally processed by 0.3 A_{260} nm microsomal membranes.

Although the endogenous level of mRNA is lost in reticulocyte lysates, it may be further reduced in order to maximize the dependence of translation on the addition of exogenous mRNA. This reduction is achieved by treatment with a calcium-activated nuclease that is thereafter inactivated by the addition of EGTA *(4)*. The system is thus somewhat disrupted with respect to the in vivo situation and is particularly sensitive to the presence of calcium ions. The resulting lysate, however, is the most efficient in vitro translation system with respect to the exogenous mRNA-stimulated incorporation of radioactive amino acids into translation products. It is therefore particularly appropriate for the study of translation products. The system is sensitive, however, to regulation by a number of factors, including hemin, double-stranded RNA, and depletion of certain metabolites. The effects of these factors on regulation of translation of various mRNAs may therefore be investigated. Despite the efficiency of reticulocyte lysates, the competition for initiation of translation by various mRNA species may differ from the in vivo situation. Products therefore may not be synthesized at in vivo proportions; the wheat germ extract cell-free system reflects the in vivo situation more faithfully. Nuclease-treated rabbit reticulocyte lysate cell-free systems are available as kits from a number of commercial suppliers.

2. Materials

All in vitro translation components are stored at −70°C. Lysates, microsomal membranes, and [^{35}S]-labeled amino acids are particularly temperature-labile and therefore should be stored in convenient aliquots at −70°C; freezing and thawing cycles must be minimized. Solutions are quick-frozen on dry ice or in liquid nitrogen prior to storage.

1. Folic acid: 1 mg/mL folic acid, 0.1 mg/mL vitamin B$_{12}$, 0.9% (w/v) NaCl, pH 7.0; filtered through a 0.45-µm filter and stored in aliquots at −20°C.
2. 2.5% (w/v) phenylhydrazine, 0.9% (w/v) sodium bicarbonate, pH 7.0 (with NaOH). Stored no longer than 1 wk at −20°C in single dose aliquots. Thawed unused solution must be discarded. Hydrazine degrades to darken the straw color.
3. Physiological saline: 0.14M NaCl, 1.5 mM MgCl$_2$, 5 mM KCl. Stored at 4°C.
4. 1 mM hemin.
5. 0.1M CaCl.
6. 7500 U/mL micrococcal nuclease in sterile distilled water. Stored at −20°C.
7. Rabbit reticulocyte lysate: This is prepared essentially as described by Pelham and Jackson *(4)*. Rabbits are made anemic by intramuscular injection of 1 mL folic acid solution on d 1, followed by six daily injections of 0.25 mL/kg body weight of 2.5% phenylhydrazine solution *(see* **Note 1**). At a reticulocyte count of at least 80%, blood is collected on d 7 or 8 by cardiac puncture into a 200-mL centrifuge tube containing approx 3000 U of heparin, and mixed well. Preparation should continue at 24°C.
   a. Blood is centrifuged at 120g, 12 min, 2°C, and plasma removed by aspiration.
   b. Cells are resuspended in 150 mL *ice cold* saline and washed at 650g for 5 min. Washing is repeated three times.
   c. The final pellets are rotated gently in the bottle, then transferred to Corex tubes (which are only half-filled). An equal volume of saline is added, the cells gently suspended, then pelleted at 1020g for 15 min at 2°C. The leukocytes (buffy coat) are then removed by aspiration with a vacuum pump.
   d. In an ice bath, an equal volume of ice-cold sterile deionized distilled water is added and the cells lysed by vigorous vortexing for 30 s *(see* **Note 2**). The suspension is then immediately centrifuged at 16,000g for 18 min at 2°C.
   e. At 4°C the supernatant is carefully removed from the pellet of membranes and cell debris. This lysate is then quick frozen in liquid nitrogen in aliquots of approx 0.5 mL.

The optimum hemin concentration is determined by varying its concentration from 0–1000 μM during the micrococcal nuclease digestion. Lysate (477.5 μL), 5 μL of 0.1M CaCl$_2$, and 5 μL of nuclease (75 U/mL final concentration) is mixed. A 97.5 μL volume of this is incubated with 2.5 μL of the relevant hemin concentration at 20°C for 20 min. A 4 μL 0.05M solution of EGTA is added to stop the digestion (*see* **Note 3**). The optimum hemin concentration is that allowing the greatest translational activity (incorporation of radioactive amino acids) in a standard cell-free incubation (*see* **Subheading 3.**). A quantity of 25 μM is generally used to ensure efficient chain initiation.

Lysates are extremely sensitive to ethanol, detergents, metals, and salts, particularly calcium. Stored at −70°C, reticulocyte lysates remain active for more than 6 mo.

8. L-[^{3}H]- or L-[^{35}S]-amino acids. A radioactive amino acid, labeled to a high specific activity (140 Ci/mmol tritiated, or approx 1 Ci/mmol [^{35}S]-labeled amino acids), is added to the translation incubation to enable detection of the translation products. An amino acid known to be abundant in the protein of interest is chosen. Radioactive solutions should preferably be aqueous; those of low pH should be neutralized with NaOH; ethanol should be removed by lyophilization, and the effect of solvents on lysate activity should be tested. [^{35}S] degrades rapidly to sulfoxide and should be aliquoted and stored at −70°C to prevent interference by sulfoxides.

9. Messenger RNA. Total RNA may be extracted from various tissues by a number of methods (*see* Chapter 11). mRNA stored in sterile dH$_2$O at −70°C is stable for more than a year. Contamination by ions, metals, and detergents should be avoided.

    Phenol may be removed by chloroform:butanol (4:1) extractions; salts are removed by precipitation of RNA in 0.4M potassium acetate, pH 6.0, in ethanol. Ethanol should be removed by lyophilization. Convenient stock concentrations for translation are 1.5 mg/mL total RNA or 150 μg/mL polyA$^+$ RNA.

10. Translation cocktail: 250 mM HEPES, pH 7.2, 400 mM KCl, 19 amino acids at 500 mM each (excluding the radioactive amino acid), 100 mM creatine phosphate.

11. 20 mM magnesium acetate, pH 7.2.

12. 2.0M potassium acetate, pH 7.2.

13. Sterile distilled H$_2$O.

Sterile techniques are used; RNase contamination is avoided by heat-treatment of glassware (250°C, 12 h) or by treatment of heat-sensitive materials with diethylpyrocarbonate, followed by rinsing in distilled water. Sterile gloves are worn throughout the procedure.

3. Method

In vitro translation procedures are best carried out in autoclaved plastic microfuge tubes (1.5 mL); a dry incubator is preferable to waterbaths for provision of a constant temperature. All preparations are performed on ice.

1. Prepare (on ice) the following reaction mix (per inculcation): 0.7 μL of dH$_2$O, 1.3 μL of 2.0M potassium acetate (*see* **Note 4**), 5 μL (10–50 μCi) of radioactive amino acid (*see* **Note 5**), and 3 μL of translation cocktail. Components are added in the

above order, vortexed, and 10 mL is aliquoted per incubation tube on ice
(*see* **Note 6**).

2. Add 10 µL (300 mg) of total mRNA (*see* **Notes 7** and **8**). A control incubation
without the addition of exogenous mRNA detects translation products of residual
endogenous reticulocyte mRNA.

3. A 10 µL volume of lysate is added last to initiate translation. If required, 0.5 A_{260} nm
U of microsomal membranes are also added at this point for cotranslational processing of translation products (*see* **Notes 9** and **10**).

4. The mixture is vortexed gently prior to incubation at 37°C for 60 min. The
reaction is stopped by placing the tubes on ice (*see* **Note 6**).

5. Detection of mRNA-directed incorporation of radioactive amino acids into
translation products is performed by determination of acid-precipitable counts.
   At the initiation and termination of the incubation, 5-µL aliquots are spotted
onto glass fiber filters that are then air-dried. Filters are then placed into 10 mL/
filter of the following solutions:
   a. 10% (v/v) cold trichloroacetic acid (TCA) for 10 min on ice.
   b. 5% (v/v) boiling TCA for 15 min, to degrade primed tRNAs.
   c. 5% (v/v) cold TCA for 10 min on ice.
   The filters are then washed in 95% (v/v) ethanol, then in 50% (v/v) ethanol-50%
(v/v) acetone, and finally in 100% (v/v) acetone. The filters are dried at 80°C for 30
min. TCA-precipitated radioactivity is determined by immersing the filters in 5 mL
of toluene-based scintillation fluid and counting in a scintillation counter.
   Exogenous mRNA-stimulated translation can be expected to result in a five-
to 30-fold increase over background of incorporation of [^{3}H]- or [^{35}S]-labeled
amino acids, respectively, into translation products.

6. An equal volume of 2% (w/v) SDS, 20% (w/v) glycerol, 0.02% (w/v) bromophenol
blue, $1M$ urea is added to the remaining 20 µL of translation mixture. This is
made $0.1M$ with respect to dithiothreitol, heated at 95°C for 6 min, and slowly
cooled to room temperature prior to loading onto a polyacrylamide gel of
appropriate concentration (between 6 and 17%). After electrophoresis, radioactive
areas of the gel are visualized by fluorography (**Fig. 1**).

4. Notes

1. Maximum anemia may be achieved by reducing the dose of phenylhydrazine on d 3,
then increasing it on following days. The reticulocyte count is determined as follows:
   a. 100 µL of blood is collected in 20 µL of 0.1% heparin in saline.
   b. 50 µL of blood heparin is incubated at 37°C for 20 min with 50 µL of 1%
   (w/v) brilliant cresyl blue, 0.6% (w/v) sodium citrate, 0.7% (w/v) sodium chloride.
   c. Reticulocytes appear under the microscope as large, round, and with blue
   granules. Erythrocytes are small, oval, and agranular.

2. The volume of water (in mL) required to lyse the reticulocyte preparation is equal
to the weight of the pellet in the tube.

3. Endogenous mRNAs of lysates are degraded by a calcium-activated nuclease
that is inactivated by EGTA. Lysates are therefore sensitive to calcium ions, the

addition of which must be avoided to prevent degradation of added mRNAs by this activated nuclease.

4. Optimum potassium concentrations may vary from 30–100 mM depending on mRNAs used and should be determined prior to definitive translations. Similarly specific mRNAs may require altered magnesium concentrations, although a concentration of 0.6–1.0 mM is generally used.

5. Specific activities greater than those described in **Subheading 2.** may result in depletion of the amino acid concerned, with subsequent inhibition of translation

6. Vigorous vortexing decreases efficiency of translation, therefore do so gently when preparing the reaction mix.

7. The optimum mRNA concentration should be determined prior to definitive experiments by varying the mRNA concentrations while keeping other variables constant. Care should be taken to avoid excess mRNA; polyadenylated RNA in excess of 1 µg has been noted to inhibit translation.

8. Heating of mRNA at 70–80°C for 1 min followed by quick cooling in an ice bath, prior to addition to the incubation mixture, has been shown to increase the efficiency of translation of GC-rich mRNA; for example, heating elastin mRNA at 70–80°C prior to translation resulted in a 100% increase, compared with unheated mRNA, of incorporation of radioactivity into translation products *(12)*.

9. Cotranslational processing of translation products may be detected by the addition of dog pancreas microsomal membranes to the translation incubation. These may be prepared as described by Jackson and Blobel *(11)* or may be ordered with a commercial translation kit. Microsomal membranes should be stored in aliquots of approx 5 A$_{260}$ nm U in 20 mM HEPES, pH 7.5, at –70°C. Repeated freezing and thawing must be avoided.

10. The addition of spermidine at approx 0.4 mM has been noted to increase translation efficiency in certain cases *(12)*, possibly by stabilizing relevant nucleic acids. However, this effect may also be lysate-dependent and should be optimized if necessary for individual lysate preparations.

References

1. Lodish, H. F. (1976) Translational control of protein synthesis. *Annu. Rev. Biochem.* **45,** 39–72.
2. McDowell, M. J., Joklik, W. K., Villa-Komaroff, L., and Lodish, H. F. (1972) Translation of reovirus messenger RNAs synthesized in vitro into reovirus polypeptides be several mammalian cell-free extracts. *Proc. Natl. Acad. Sci. USA*
3. Sampson, J., Mathews, M. B., Osborn, M., and Borghetti, A. F. (1972) Hemoglobin messenger ribonucleic acid translation in cell-free systems from rat and mouse liver and Landschutz ascites cells. *Biochemistry* **11,** 3636–3640.
4. Pelham, H. R. B. and Jackson, R. J. (1976) An efficient mRNA-dependent translation system from reticulocyte lysates. *Eur. J. Biochem.* **67,** 247–256.
5. Carlier, A. R. and Peumans, W. J. (1976) The rye embryo system as an alternative to the wheat-system for protein synthesis in vitro. *Biochem. Biophys. Acta* **447,** 436–444.

6. Roberts, B. E. and Paterson, B. M. (1973) Efficient translation of tobacco mosaic virus RNA and rabbit globin 9S RNA in a cell-free system from commercial wheat germ. *Proc. Natl. Acad. Sci. USA* **70,** 2330–2334.
7. Kessler, S. W. (1981) Use of protein A-bearing staphylococci for the immunoprecipitation and isolation of antigens from cells, in *Methods in Enzymology* (Langone, J. J. and Van Vunakis, H., eds.), Academic, New York, pp. 441–459.
8. Laemmli, U. K. (1970) Cleavage of structural proteins during the assembly of the head of bacteriophage T4. *Nature* **227,** 680–685.
9. Banner, W. M. and Laskey, R. A. (1974) A film detection method for tritium-labeled proteins and nucleic acids in polyacrylamide gels. *Eur. J. Biochem.* **46,** 83–88.
10. Shields, D. and Blobel, G. (1978) Efficient cleavage and segregation of nascent presecretory proteins in a reticulocyte lysate supplemented with microsomal membranes. *J. Biol. Chem.* **253,** 3753–3706.
11. Jackson, R. C. and Blobel, G. (1977) Post-translational cleavage of presecretory proteins with an extract of rough microsomes, from dog pancreas, with signal peptidase activity. *Proc. Natl. Acad. Sci. USA* **74,** 5598–5602.
12. Karr, S. R., Rich, C. B., Foster, J. A., and Przybyla, A. (1981) Optimum conditions for cell-free synthesis of elastin. *Coll. Res.* **1,** 73–81.

31

In Vitro Translation of Messenger RNA in a Wheat Germ Extract Cell-Free System

Louise Olliver, Anne Grobler-Rabie, and Charles D. Boyd

1. Introduction

The wheat germ extract in vitro translation system has been used widely for faithful and efficient translation of viral and eukaryotic messenger RNAs in a heterologous cell-free system *(1–9)*. With respect to the yield of translation products, the wheat germ extract is less efficient than most reticulocyte lysate cell-free systems (*see* Chapters 29, 30, and 32). There are advantages, however, of using wheat germ extracts:

1. The in vivo competition of mRNAs for translation is more accurately represented, making the wheat germ system preferable for studying regulation of translation *(1)*.
2. Particularly low levels of endogenous mRNA and the endogenous nuclease activity *(10)* obviate the requirement for treatment with a calcium-activated nuclease. There is, therefore, less disruption of the in vivo situation and contamination with calcium ions is less harmful. The identification of all sizes of exogenous mRNA-directed translation products is facilitated because of the low levels of endogenous mRNA present.
3. There is no posttranslational modification of translation products; primary products are therefore investigated, although processing may be achieved by the addition of microsomal membranes to the translation reaction.
4. The ionic conditions of the reaction may be altered to optimize the translation of large or small RNAs *(2)* (*see* **Note 1**).

Translational activity is optimized by the incorporation of an energy-generating system of ATP, GTP, creatine phosphate, and creatine kinase *(3)*. Wheat germ is inexpensive and commercially available (*see* **Note 2**); preparation

From: *Methods in Molecular Biology, Vol. 86: RNA Isolation and Characterization Protocols*
Edited by: R. Rapley and D. L. Manning © Humana Press Inc., Totowa, NJ

of the extract is rapid and simple, resulting in high yields. Wheat germ extract cell-free system kits are also commercially available.

2. Materials

Components of the wheat germ in vitro translation system are heat-labile and must be stored in aliquots of convenient volumes at −70°C. Freeze-thaw cycles must be minimized. Sterile techniques are used throughout. RNase contamination is prevented by heat-sterilization (250°C, 8 h) of glassware and tips, and so on, or by diethyl pyrocarbonate treatment of glassware, followed by thorough rinsing of equipment in sterile distilled water.

1. Wheat germ extract: This is prepared essentially as described by Roberts and Paterson *(4)*. The procedure must be carried out at 4°C, preferably in plastic containers since initiation factors stick to glass. Fresh wheat germ (approx 5 g) (*see* **Note 2**) is ground with an equal weight of sand and 28 mL of 20 mM HEPES, pH 7.6, 100 mM KCl, 1 mM magnesium acetate, 2 mM CaCl$_2$, and 6 mM 2-mercapteothanol, added gradually. This mixture is then centrifuged at 28,000g for 10 min at 2°C, pH 6.5. This pH prevents the release of endogenous mRNA from polysomes and therefore removes the requirement for a preincubation to allow polysome formation *(4,5)*. The supernatant (S-28) is then separated from endogenous amino acids and plant pigments that are inhibitory to translation, by chromatography through Sephadex G-25 (coarse) in 20 mM HEPES, pH 7.6, 120 mM KCl, 5 mM magnesium acetate, and 6 mM 2-mercaptoethanol. Reverse chromatography will prevent the loss of amino acids. Fractions of more than 20 A_{260} nm/mL are pooled before being stored in aliquots at a concentration of approx 100 A_{260} nm/mL, at −70°C. The extract remains translationally active for a year or more.

2. L-[^{3}H]- or L-[^{35}S]-amino acids: 10–50 μCi of an appropriate amino acid (abundant in the protein[s] of interest) is added to the reaction to allow detection of translation products. Convenient specific activities are 140 Ci/mmol tritiated, or 1 Ci/mmol [^{35}S]-amino acids, respectively (*see* **Note 3**). Aqueous solutions should be used since ethanol, salts, detergents, and various solvents interfere with translation. Ethanol should be removed by lyophilization and the effects on translation of other solutions should be determined prior to their use. [^{35}S]-labeled amino acids must be stored in small aliquots at −70°C where they remain stable for up to 6 mo, after which time sulfoxide products of degradation inhibit translation.

3. Messenger RNA: The extraction of both total and polyadenylated RNA has been described by a number of authors *(10–12)*. Total RNA (1.5 mg/mL) or 150 μg/mL polyadenylated RNA (in sterile distilled water) are convenient stock concentrations. RNA is stable for more than a year at −70°C. Contamination with potassium (*see* **Note 1**), phenol, and ethanol must be prevented by 70% (v/v) ethanol washes, chloroform:butanol (4:1) extractions, and lyophilization respectively.

4. 10X energy mix: 10 mM ATP, 200 μM GTP, 80 mM creatine phosphate. Potassium salts of the nucleotide triphosphates should be used and the final pH adjusted (if necessary) to 7.4–7.6 with sodium hydroxide.

5. 0.5–1.0M potassium acetate (*see* **Note 1**), 25 mM magnesium acetate.
6. 20 mM dithiothreitol.
7. 0.6–1.2 mM spermine or 4.0–8.0 mM spermidine (*see* **Note 4**).
8. 0.2M HEPES, pH 7.4–7.6 (*see* **Note 5**).
9. 200–500 μg/mL creatine kinase (*see* **Note 6**).

3. Method

All preparations are carried out on ice. After use, components are quick-frozen on dry ice. Reactions are carried out in sterile plastic microfuge tubes.

1. Mix the following solutions (all components are v/50 μL): 5 μL of energy mix, 5 μL of potassium and magnesium acetate, 5 μL of dithiothreitol, 5 μL of HEPES, 5 μL of spermine, 10 μL of 0.3–8.0 μg mRNA in dH$_2$O, (*see* **Note 7**), 10 μL of wheat germ extract, 10 μL of creatine kinase (0.8–1.0 A$_{260}$ U), and 5 μL of creatine kinase. If a number of incubations are to be made, a master mix of the first five solutions may be prepared and 25 μL aliquoted into each reaction tube. Creatine kinase is added last to ensure that no energy is wasted. The solutions are mixed by tapping the tube or by gentle vortexing. Microsomal membranes (0.5 A$_{260}$ U) may be added before the creatine kinase to detect cotranslational modification of translation products (*see* **Note 8**).
2. Incubate at 28°C for 1 h (*see* **Note 9**). The reaction is terminated by placing the tubes at 4°C.
3. Incorporation of radioactive amino acids into mRNA-derived translation products is detected by TCA precipitation of an aliquot of the reaction (*see* **Chapter 30** and **Note 10**). Incorporation of radioactivity into translation products is generally not as well-stimulated by mRNA added to wheat germ extracts as it is in described reticulocyte lysates.
4. The remaining in vitro translation products may be analyzed further by standard techniques, including tryptic mapping and ion-exchange chromatography, but specific products may be analyzed by immunoprecipitation followed by SDS-polyacrylamide gel electrophoresis.

4. Notes

1. Wheat germ extract translational activity is particularly sensitive to variation in the concentration of potassium ions. At concentrations lower than 70 mM, small mRNAs are preferentially translated, whereas larger mRNAs are translated at potassium acetate concentrations of 70 mM or greater *(2,5)*. Polypeptides of up to 200 kDa are synthesized under correct ionic conditions *(9)*. Furthermore, chloride ions appear to inhibit translation such that potassium acetate should preferably be used *(5)*. In this context, residual potassium should be removed from RNA preparations, by 70% (v/v) ethanol washes.
2. Inherent translational activity varies with the batch of wheat germ. Israeli mills (for example "Bar-Rav" Mill, Tel Aviv) supply wheat germ, the extracts of which are usually active.

3. Most of the endogenous amino acids are removed by chromatography through Sephadex G-25 (coarse) (*see* Chapter 26). Depending on the batch of wheat germ extract, addition of amino acids (to 25 μ*M*) and/or tRNA (to 58 μg/mL) may be necessary to optimize translational activity. Wheat germ extract is particularly sensitive to amino acid starvation; use of radioactive amino acids at specific activities greater than those suggested may result in inhibition of translation because of amino acid starvation.

4. The use of either spermine or spermidine generally stimulates translation, and is essential for the synthesis of larger polypeptides *(5)*, probably by stabilizing longer mRNAs. Omission of either compound will increase the optimum magnesium acetate concentration to 4.0–4.3 m*M*.

5. HEPES has been shown to buffer the wheat germ extract in vitro translation system more effectively than Tris-acetate *(4)*. Use of the latter will alter the optimum potassium and magnesium concentration.

6. Commercial preparations of creatine kinase differ with respect to the levels of nuclease contamination. This must be considered when larger amounts of the enzyme are to be used.

7. Heating of large mRNAs at 70°C for 1 min followed by rapid cooling on ice increases the efficiency of their translation in wheat germ extract in vitro translation systems.

8. Cotranslational processing of translation products may be detected by the addition of dog pancreas microsomal membranes to the translation incubation. They may be prepared as described by Jackson and Blobel *(12)* or may be ordered with a commercial translation kit. Microsomal membranes should be stored in aliquots of approx 5 A_{260} nm U in 20 m*M* HEPES, pH 7.5, at –70°C. Repeated freezing and thawing must be avoided.

9. mRNA-stimulated incorporation of radioactive amino acids into translation products is linear, after a 5 min lag, for 50 min and is complete after 90 min. The system is labile at temperatures >30°C; optimum activity is achieved at 25–30°C depending on the batch of wheat germ extract. An incubation temperature of 28°C is generally used.

10. In order to obtain maximum translational activity, it is necessary to determine the optima for each preparation of wheat germ extract; mRNA concentration, potassium and magnesium concentrations, and incubation temperature. Take into account the concentration of salts in the wheat germ extract column eluate.

References

1. Steward, A. G., Lloyd, M., and Arnstein, H. R. V. (1977) Maintenance of the ratio of α and β globin synthesis in rabbit reticulocytes. *Eur. J. Biochem.* **80,** 453–459.
2. Benveniste, K., Wilczek, J., Ruggieri, A., and Stern, R. (1976) Translation of collagen messenger RNA in a system derived from wheat germ. *Biochemistry* **15,** 830–835.
3. Huntner, A. R., Farrell, P. J., Jackson, R. J., and Hunt, T. (1977) The role of polyamines in cell-free protein in the wheat germ system. *Eur. J. Biochem.* **75,** 149–157.

4. Roberts, B. E. and Paterson, B. M. (1973) Efficient translation of tobacco mosaic virus RNA and rabbit globin 9S RNA in a cell-free system from commercial wheat germ. *Proc. Natl. Acad. Sci. USA* **70**, 2330–2334.

5. Davies, J. W., Aalbers, A. M. J., Stuik, E. J., and van Kammen, A. (1977) Translation of cowpea mosaic RNA in cell-free extract from wheat germ. *FEBS Lett.* **77**, 265–269.

6. Boedtker, H., Frischauf, A. M., and Lehrach, H. (1976) Isolation and translation of calvaria procollagen messenger ribonucleic acids. *Biochemistry* **15**, 4765–4770.

7. Patrinou-Georgoulas, M. and John, H. A. (1977) The genes and mRNA coding for the theory chains of chick embryonic skeletal myosin. *Cell* **12**, 491–499.

8. Larkins, B. A., Jones, R. A., and Tsai, C. Y. (1976) Isolation and in vitro translation of zein messenger ribonucleic acid. *Biochemistry* **15**, 5506–5511.

9. Schroder, J., Betz, B., and Hahlbrock, K. (1976) Light-induced enzyme synthesis in cell suspension cultures of *petroselinum. Eur. J. Biochem.* **67**, 527–541.

10. Pelham, H. R. B. and Jackson, R. J. (1976) An efficient mRNA-dependent translation system from reticuloctye lysates. *Eur. J. Biochem.* **67**, 247–256.

11. Darnbrough, C. H., Legon, S., Hunt, T., and Jackson, R. J. (1973) Initiation of protein synthesis: evidence for messenger RNA-independent binding of methionyl-transfer RNA to the 40S ribosomal subunit. *J. Mol. Biol.* **76**, 379–403.

12. Jackson, R. C. and Blobel, G. (1977) Post-translational cleavage of presecretory proteins with an extract of rough microsomes, from dog pancreas, with signal peptidase activity. *Proc. Natl. Acad. Sci. USA* **74**, 5598–5602.

32

The *Xenopus* Egg Extract
Translation System

Glenn M. Matthews and Alan Colman

1. Introduction

A full analysis of the post-translational modifications that a given protein undergoes during transit through the secretory pathway may, in some cases, only be performed by analysis of the natural protein, expressed in its normal tissue.

Often this is not possible since the quantity produced is too small to give strong signals after, for example, radiolabeling and immunoprecipitation. Expression of cDNA clones in *Xenopus* oocytes or cultured cells has been widely used to determine not only the nature of covalent modifications, but also the fate, that is, whether membrane bound, secreted, or resident in the secretory pathway, of a wide range of proteins.

A major barrier to analyses performed by expression in living cells is that some method is required to visualize the protein under investigation, normally by means of an antibody capable of being used for immunoprecipitation. Often a useful antibody may be difficult to produce or, during the early stages of analysis of a newly cloned sequence, it may be desirable to verify the secretory phenotype and characterize the primary posttranslational modifications that occur to the protein before raising antisera. It is at this stage where cell-free systems can be most useful.

The "traditional" translocating in vitro systems, where reticulocyte lysate (*see* Chapter 30) or wheat-germ extracts (*see* Chapter 31) are combined with canine pancreatic membranes, solve the problem of background translation, and have been enormously useful in the development of current knowledge of the mechanisms of translocation and early processing events. For routine analysis, however, they can often be difficult to use, since the translocation capac-

From: *Methods in Molecular Biology, Vol. 86: RNA Isolation and Characterization Protocols*
Edited by: R. Rapley and D. L. Manning © Humana Press Inc., Totowa, NJ

ity can be limited and the membranes are often quite fragile, making protease protection and fractionation experiments difficult.

The *Xenopus* egg extract *(1)* has a high capacity for translocation, signal sequence cleavage, and *N*-glycosylation, and so gives consistent processing patterns across a wide range of added mRNA concentrations, whereas the stability of the membranes present allows the use of sucrose gradient fractionation and protease protection to verify the location of translation products. The ability of the extract to support assembly of multimeric proteins and perform, to a limited extent, *O*-glycosylation and mannose-6-phosphorylation further extend its utility as an analytical tool.

Preparation of the extract, which is based on the method described by Murray *(2)* for the preparation of extracts for cell-cycle studies, involves centrifugal lysis of *Xenopus* eggs. In this crude form, with only ^{35}S-methionine added, a translation reaction produces approx 60 μg/mL of protein from the endogenous mRNA. This figure is increased to 140 μg/mL on addition of creatine phosphate to 7 mM and 10% by volume of an S-100 fraction of rabbit reticulocyte lysate. After ribonuclease treatment to remove endogenous mRNAs, approximately half of this activity can be restored by the addition of poly A$^+$ mRNA. When a single synthetic mRNA, encoding a secretory protein, is used to program translation, yields in excess of 10 μg/mL of translocated and processed product can be obtained. The extract can be frozen for storage, allowing many independent experiments to be performed on a single batch. After freezing, however, the addition of creatine phosphate is necessary for activity, whereas reticulocyte lysate S-100 and 1 mM spermidine stimulate translation. This combination restores 25–60% of original activity. When performed in the presence of 800 μCi/mL of ^{35}S-methionine, this range of activities allows most translation products to be easily detected by overnight fluorography of SDS gels.

2. Materials

1. Frogs: *Xenopus* stock.
2. High-Salt Modified Barth's X (MBS) It is supplemented by addition of 1.28 g of NaCl/L to give a final concentration of 110 mM.
3. Folligon (serum gonadotrophin) and Chorulon (chorionic gonadotrophin) are obtained from Intervet (Cambridge, UK) and dissolved in the solvent provided.
4. Dejellying solution: 2% cysteine HCl titrated to pH 7.7 with NaOH.
5. Extraction buffer: 100 mM KCl, 0.1 mM CaCl$_2$, 1 mM MgCl$_2$, 50 mM sucrose, and 10 mM HEPES-KOH, pH 7.7 (titrated at 10 mM).
6. Versilube VF50 (General Electric): This oil has a density between that of the eggs and of the extraction buffer.
7. Cytochalasin B (Sigma, Poole, UK): Stock at 10 mg/mL in DMSO; store at 4°C

8. Aprotinin (Boehringer, Lewes, UK): Stock at 10 mg/mL in water; store at −20°C.

9. RNase A (Boehringer): Stock at 1 mg/mL in water; store in aliquots at −20°C.

10. Ribonuclease inhibitor (Boehringer): This is normally supplied at 50 U/μL.

11. Dithiothreitol, Boehringer (DTT): Stock at $1M$; stored in aliquots at −20°C. This is freshly diluted to 100 mM before addition to the extract.

12. tRNA (calf liver, Boehringer): Stock at 5 mg/mL in water; store at −20°C in aliquots.

13. Creatine phosphate (Boehringer): Stock at 350 mM in water; store at −20°C in small aliquots.

14. [^{35}S]-Methionine (SJ 204, Amersham, Little Chalfont, UK): Divide into 100-μCi aliquots on first thawing, and store at −70°C.

15. Spermidine (Sigma): Stock at 120 mM; store at −20°C.

16. Rabbit reticulocyte lysate/S-100 extract: It is only necessary to prepare the S-100 fraction if it is desirable to exclude exogenous ribosomes from the reaction. Whole (nuclease-treated) reticulocyte lysate can be added in place of the S-100 fraction if this is not important. We normally use the reticulocyte lysate supplied by Bethesda Research Laboratories (Paisley, UK), but nuclease-treated material from any source would probably be as effective. To prepare an S-100 fraction, 100-μL portions of reticulocyte lysate are centrifuged at 100,000g in a TLA-100 rotor for 2 h, 80 μL of the supernatant is recovered, taking care to avoid the ribosomal pellet, and flash-frozen in liquid nitrogen as 10-μL aliquots, before storage at −70°C.

17. Triton X-100 (Surfact-Amps X-100, Pierce, Chester, UK): This is supplied as a 10% solution in 10-mL vials. After opening, unused material can be stored in the dark at 4°C.

18. Phenylmethylsulfonyl fluoride (PMSF, Sigma): Stock at 100 mM in propan-2-ol. Store at 4°C. Aqueous solutions of PMSF lose activity very rapidly, and should be made up immediately before use.

19. 1% Triton X-100, 1 mM PMSF: Freshly made from the stock solutions above.

20. 2X T buffer: 100 mM KCl, 10 mM Mg acetate, 200 mM NaCl, and 40 mM Tris-HCl, pH 7.6. Sterilize by filtration and store at −20°C. This is blended with sucrose from a 40% stock solution to generate 1X T + 10% and 1X T + 20% sucrose solutions. Unused portions of T + 10 and T + 20 can be stored at −20°C.

21. Proteinase K (Boehringer): Stock solutions at 25 mg/mL in autoclaved 50% glycerol appear to be stable on storage at −20°C. Before use in protease protection experiments, the stock is diluted to 1 mg/mL in 10% sucrose and allowed to stand at room temperature for 15 min to digest any contaminating enzymes, such as lipases.

22. Sodium carbonate: Prepare a fresh $1M$ stock for each alkaline sucrose gradient experiment. Check that the pH of a 100-mM solution is 11. This can then be diluted (1 in 5 with water) to 200 mM for treatment of membrane fractions and can be blended with 40% sucrose (1 vol $1M$ sodium carbonate, 4 vol water, and 5 vol 40% sucrose) to form the alkaline 20% sucrose cushion.

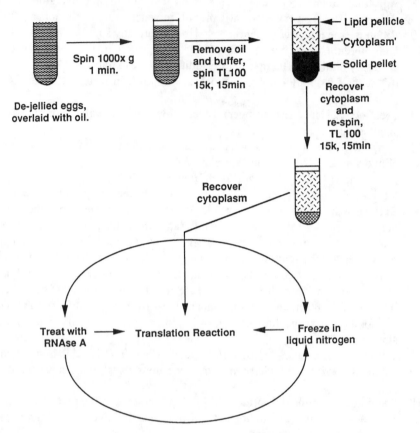

Fig. 1. Stages in preparation of an extract.

23. 1M HCl.
24. Acetyl-Asn-Tyr-Thr-amide (custom synthesized by Alta Bioscience, University of Birmingham): This is relatively insoluble in water, so a 100-mM stock solution must be prepared in DMSO. Dilutions from this can then be made in water.
25. Centrifuges: We routinely use a Beckman TL-100 for the crushing spin, which has the advantage of efficient refrigeration, an appropriate tube/rotor format, and rapid acceleration to the set speed. Conditions specified below, therefore, all relate to this model. Many other machines could probably be used instead, but the only alternative we have tried is an Eppendorf model 5414 microfuge in a 4°C cold room. This gave good results, but the yield was compromised because of the use of a fixed-angle rotor. For the low-speed spin before the crushing step, almost any refrigerated bench-top model with a swing-out rotor should be adequate.
26. Cold room: Extracts prepared in a 4°C cold room are invariably more active than those exposed to room temperature during preparation.

3. Methods

3.1. Preparation of the Extract (See Notes 1–7)

3.1.1. Preparation of the Basic Extract

A flow diagram indicating the stages of the preparation of the extract is shown in **Fig. 1**. Before starting to make an extract, ensure that all buffers, tubes, and rotors are cooled to 4°C and that all necessary materials are at hand, since the faster the whole procedure can be performed, the better the final extract. Unless a temperature is otherwise stated, all the procedures, after the dejellying step, should be performed on ice and, ideally, in a cold room.

1. Large adult female *Xenopus laevis* are primed by injection with 50–100 U of Folligon, on d 1. Three to five days later, they are induced to lay by injection of 500–750 U of Chorulon, late in the evening of the day before the extract is to be made. Frogs are left overnight to shed eggs into high-salt MBS, which prevents activation.
2. Transfer approx 30 mL of loosely packed eggs to a 250-mL glass beaker, rinse a few times with high-salt MBS to remove debris, and aspirate off any obviously dead eggs with a Pasteur pipet. Pour away as much supernatant buffer as possible, and add 100 mL of dejellying solution, repeat this two or three times. Agitate the suspension of eggs occasionally by swirling for 5–10 min. The dissolution of the jelly coats can be easily seen as a marked reduction in the volume occupied by the eggs. When this has occurred, rinse once more in dejellying solution and then transfer, by multiple washes, to ice cold extraction buffer.
3. Fill four 2-mL polyallomer TLS 55 centrifuge tubes by gently pipeting the eggs using a wide-bore Pasteur pipet, transferring as little buffer as possible. Allow the eggs to settle for a minute or so, then carefully remove the supernatant buffer, top the tubes up with Versilube VF50, and centrifuge at 500g for 1 min at 4°C in a swing-out rotor. Depending on the rotor format, it may be useful to place the tubes inside another container, e.g., precooled 7-mL bijoux tubes, for easier handling, thermal insulation during transfer, and to contain any spillage. The eggs should pack tightly, but not lyse during this step, and the buffer should be separated from the eggs by a layer of oil.
4. Remove the supernatant buffer and then the oil by careful aspiration with a 200-µL micropipet, load the tubes into a TLS 55 rotor, and crush the eggs by centrifugation at 20,000g for 15 min at 4°C. This produces a multilayered lysate of which the desired product, the viscous amber middle layer normally referred to as cytoplasm, constitutes around 40% by volume. Recover this layer by inserting a Pasteur pipet through the lipid pellicle.
5. Pool the product from all tubes, estimate the volume, and add 5 µL/mL of 10 mg/mL cytochalasin B (mL). Mix by gentle pipeting, transfer to fresh polycarbonate 1.5 mL TLS 55 centrifuge tubes, and centrifuge again at 20,000g for 15 min at 4°C. The cytoplasm should now occupy most of the volume, but this must be removed very slowly to avoid disturbing the pellet.

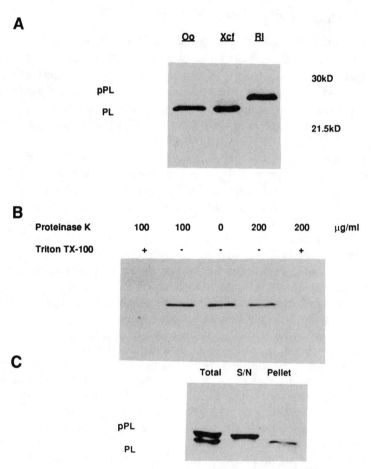

Fig. 2. Translation, protease protection, and neutral sucrose gradient fractionation of bovine prolactin. **(A)** Translation products obtained from synthetic prolactin mRNA translated in *Xenopus* oocytes (Oo), the *Xenopus* egg cell-free extract (Xcf), and rabbit reticulocyte lysate (Rl). Prolactin is not *N*-glycosylated, so signal sequence cleavage of preprolactin (pPL) to prolactin can be seen as a reduction in molecular weight in the *Xenopus* systems relative to the reticulocyte lysate product. The *Xenopus* oocyte sample was immunoprecipitated before electrophoresis, whereas the in vitro products were loaded directly onto the gel. **(B)** Result of a protease protection experiment performed on a *Xenopus* egg extract translation mixture, again programmed with bovine prolactin mRNA. **(C)** Result of a neutral sucrose gradient fractionation of a *Xenopus* egg extract translation programmed with excess prolactin mRNA to saturate the translocation apparatus and provide a marker for unsegregated protein.

6. Add 1 μL of 10 mg/mL aprotinin/mL of extract, and mix gently but thoroughly.

The extract should now be ready for translation reactions, frozen immediately, or mRNA depleted as described in **Subheading 3.1.2.**

3.1.2. Depletion of Endogenous mRNAs

As with the preparation of the basic extract, this is best performed in a cold room, where the activity of the ribonuclease is more readily controlled.

1. Dilute stock RNase A to 100× the final desired concentration.
2. Add 1 μL of diluted RNase to a series of screw-capped 1.5-mL microfuge tubes, and then add 100 μL of extract, mixing thoroughly by pipeting gently. Incubate at 10°C for 15 min.
3. Transfer to ice, and add 1 μL of 100 mM DTT to each tube, followed by 50 U of ribonuclease inhibitor. Mix well, and incubate at 10°C for a further 10 min. Then add 2 μL of 5 mg/mL calf liver tRNA. This product should now be used for translation reactions or frozen in liquid nitrogen for storage as soon as possible.

3.2. The Translation Reaction (See Notes 8–14)

1. Thaw frozen extracts at room temperature until just liquid, and then place on ice.
2. Meanwhile, distribute mRNAs to be translated into 0.5- or 1.5-mL microfuge tubes on ice.
3. To each 100 μL aliquot of extract add: 10 μL reticulocyte lysate S-100, 1 μL 120 mM spermidine, 2.5 μL 350 mM creatine phosphate, and 100 μCi ^{35}S-methionine. Add aliquots (10–50 μL) of this mixture to the tubes containing mRNAs, mixing well, and then incubate at 21°C for 1 h.
4. If highly radioactive synthetic mRNA was used to program the reactions, treat, at the end of the reaction, with 10 μg/mL RNase A for 15 min at 21°C to remove the radioactive background, which the mRNA contributes to further analyses.
5. For storage, the reaction can be stopped by freezing at this stage.
6. For analysis by gel electrophoresis or TCA precipitation *(1)*, reaction products should be diluted in 4 vol of 1% Triton X-100, 1 mM PMSF before either adding an equal volume of twofold concentrated SDS-PAGE sample buffer or spotting to filters.

3.3. Analysis of Translation Products (See Notes 13–15)

Simple gel electrophoresis of translation products can often give an indication of a secretory phenotype. Signal sequence cleavage reduces molecular weight by 2–3 kDa (*see* **Fig. 2A**), whereas N-glycosylation normally reduces mobility relative to unmodified protein produced in wheat-germ extract or reticulocyte lysate (*see* **Fig. 3A**). A simple mobility shift could, however, be caused by a range of other factors, and it is normally necessary to confirm that the protein has been translocated into membranes by performing either protease protection or sucrose fractionation experiments. Further fractionation,

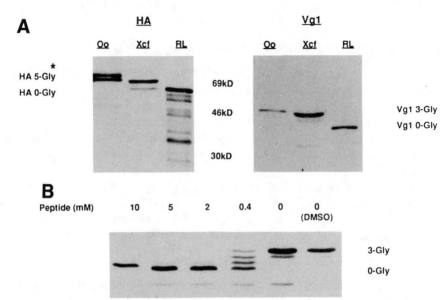

Fig. 3. Translation of N-glycosylated proteins and the partial inhibition of glycosylation. **(A)** Products obtained from the translation of *Xenopus* Vg1 and influenza virus hemagglutinin (HA) synthetic mRNAs in the *Xenopus* oocyte (Oo), *Xenopus* egg cell-free extract (Xcf), and rabbit reticulocyte lysate (Rl). The occupation of five glycosylation sites on HA (HA 5-Gly) and three on Vg1 (Vg1 3-Gly) can be clearly seen to have retarded these proteins relative to the unglycosylated reticulocyte lysate products (0-Gly). **(B)** Result of *Xenopus* cell-free extract translation reactions, programmed with synthetic mRNA encoding Vg1, performed in the presence of the indicated concentration series of the tripeptide Asn-Tyr-Thr. Total inhibition of N-glycosylation can be seen above 2 mM peptide, whereas the partial reaction, at 0.4 mM tripeptide, allows the number of N-glycosylation sites occupied to be directly determined.

on alkaline sucrose gradients, can be used to determine whether the protein is free within the lumen of the endoplasmic reticulum or integral to the membrane, whereas specific inhibition of N-glycosylation can demonstrate that this is responsible for any apparent increase in molecular weight. Examples of protease protection and neutral sucrose gradient fractionation experiments are shown in **Fig. 2B** and **C**, respectively, and inhibition of N-glycosylation is shown in **Fig. 3B**.

3.3.1. Protease Protection

1. Remove 3×10 µL aliquots from the translation reactions to be assayed, and place on ice. If it is necessary to use less than this, for example, because the

products are to be analyzed by a variety of other methods, dilute the reaction by addition of up to 4 vol of 1X T buffer + 10% sucrose.

2. Add 1 μL of 10% Triton X-100 to one of each set of tubes in order to disrupt the membranes present and thus provide a positive control for proteolysis.
3. Add 1 μL of 1 mg/mL proteinase K solution in 10% sucrose to the tube with Triton added and one of the other two (the third aliquot, with no additions, serves as a control for stability during the procedure), and incubate on ice for 1 h.
4. Freshly dilute stock 100 mM PMSF with 3 vol of 10% sucrose: Add 1 μL of this to each reaction to give a final concentration of 2–2.5 mM, and continue the incubation on ice for a further 15 min.
5. Add 100 μL of 1X SDS-PAGE sample buffer, including 1% Triton X-100, and heat in a boiling water bath for 5 min. If the sample was diluted before protease treatment, add a smaller volume of 2X SDS-PAGE buffer to ensure that the final mixture contains 10% of extract by volume.
6. Analyze by SDS-PAGE, including an untreated sample of the same reaction at an equivalent dilution, as a marker.

3.3.2. Neutral Sucrose Gradient Fractionation

Sucrose gradient fractionation can be useful not only as an analytical technique, but can also provide a significant degree of purification before, for example, performing an activity assay on the translation products. An example of the results of this method of fractionation is shown in **Fig. 2C**.

If this procedure is to be followed by the alkaline sucrose fractionation described below, it is advisable to start with a translation reaction of at least 50 μL since the volumes will be easier to handle.

1. On ice, dilute the translation reaction (by gentle pipeting) into 10 vol of 1X T buffer plus 10% sucrose. Retain a small portion as a marker for SDS-PAGE analysis, and carefully layer the remainder onto a 1-mL step of 1X T buffer plus 20% sucrose in a 1.5-mL polycarbonate centrifuge tube.
2. Centrifuge at 30,000 rpm (40,000g) in a TLS 55 rotor for 30 min at 4°C.
3. Recover the top (10% sucrose) layer, containing the cytosolic proteins, by aspiration: It is rarely necessary to recover all of this, and it is easier to avoid the dilution caused by mixing with the 20% step if only about half to three-quarters of this fraction is recovered. Remove and discard the rest of the sucrose buffer, taking care not to disturb the membrane pellet, which can be seen as a brown spot on the bottom of the tube.
4. If membrane stability is required in later analysis, such as protease protection or alkaline sucrose fractionation, gently resuspend the membrane pellet in T buffer plus 10% sucrose in a volume equivalent to half that originally loaded onto the gradient. Alternatively, dissolve the membranes in 1% Triton X-100, 1 mM PMSF.
5. Analyze equivalent portions of the total reaction and each fraction by SDS-PAGE.

3.3.3. Alkaline Sucrose Gradient Fractionation

Alkaline treatment disrupts the membrane vesicles, liberating lumenal proteins, without dissolving the lipid bilayer. After fractionation on a sucrose step gradient, the lumenal proteins remain in the supernatant, whereas membrane-bound components are pelleted.

1. To 100 μL of membranes from **Subheading 3.3.2., step 4**, add an equal volume of 200 mM Na$_2$CO$_3$. Incubate on ice for 30 min.
2. Layer onto a 250-μL step of 0.1M Na$_2$CO$_3$ in 20% sucrose in a 1.5-mL polycarbonate centrifuge tube, and centrifuge for 1 h at 100,000g at 4°C in a TLS 55 rotor.
3. Recover the supernatant (again, it is best to accept a loss here and only remove 100–150 μL), discard the 20% sucrose step, and redissolve the pellet in 1% Triton X-100, 1 mM PMSF.
4. Before analysis by SDS-PAGE, neutralize the supernatant by addition of HCl. Approximately 10% by volume of 1M HCl is required, and this should be added slowly, with mixing, to prevent local precipitation of proteins. The progress of this titration can be checked by spotting a small volume onto suitable pH indicator paper.

3.3.4. Inhibition of N-Glycosylation

Traditionally, the number of N-glycosylation sites occupied on a polypeptide chain has been measured by partial endoglycosidase H digestion. The use of the competitor tripeptide (acetyl)-Asn-Tyr-Thr-(amide) to inhibit N-glycosylation partially in the *Xenopus* extract, however, provides an alternative approach to this question.

1. Prepare a twofold dilution series of tripeptide in water, covering the range from 50–1.5 mM (6 points).
2. Assemble, on ice, a translation reaction of at least 80 μL final vol, including the mRNA for the protein under investigation.
3. Add 9 μL of the translation reaction to 1 μL of each dilution of tripeptide and to 1 μL of 50% DMSO (as a control for the effect of DMSO on the reaction). Incubate these mixtures, together with the remainder, which serves as a positive control, at 21°C for 1 h, and then analyze by SDS-PAGE, including a reticulocyte lysate translation product as a marker for unglycosylated preprotein. Partial inhibition of glycosylation will be seen at the low end of the concentration series.

4. Notes

1. As with any translation system, ribonuclease contamination should be avoided. All tubes, tips, and where possible, reagents, must be autoclaved or treated by some other means to inactivate ribonuclease. The extract is a complex mixture that, during a translation reaction, is supporting a wide range of processes. It is

therefore important to ensure that no inactivating contaminants are present in the materials added, by using the highest quality reagents available and for stock solutions to be made up in double-glass distilled or reverse-osmosis-treated water.

2. Egg quality varies from one female to another and, sometimes, with the time of year. Even the poorest eggs will give an extract that incorporates ^{35}S-methionine into proteins before ribonuclease treatment. Some batches are, however, compromised beyond this stage, particularly after freezing. A good guide to the quality of a batch of eggs is their ability to undergo fertilization. Extracts produced from eggs that do not fertilize well may benefit from the addition of sucrose to 200 mM from a 2M stock, before freezing.

3. Very occasionally, a batch of eggs will fail to lyse well. When this happens, a significant proportion (30% or more) of the extract volume will be gray, instead of amber, after the first spin. These extracts never perform well and should be abandoned at this stage.

4. There are many steps involved in generating an extract, so the chances of a batch being compromised by a faulty reagent or an error in handling are significant. Unfortunately, it is not possible to assay an extract during preparation, but performing reactions on samples at each stage of the process (i.e., fresh and frozen material both before and after RNase treatment) helps, although retrospectively, to locate any problem areas. For comparison, it is best to add spermidine, creatine phosphate, and reticulocyte lysate S-100 to all samples assayed. Extracts that have not been frozen should incorporate ^{35}S-methionine without these being present.

5. To freeze, divide the extract into aliquots of 100 µL or less in ice-cold microfuge tubes (this should already be the case if the extract has been ribonuclease-treated), and plunge into liquid nitrogen for a min. After freezing, do not allow the extract to thaw until it is to be used. Activity, relative to that of an aliquot thawed and tested immediately after freezing, is unaffected by storage at −70°C for a number of months, but storage in liquid nitrogen is preferable, if this is available.

6. To assay for recovery from freezing, an aliquot from a batch of frozen extract should simply be thawed and used to set up a translation reaction. It is tempting to economize on reagents by setting up a translation reaction with all reagents present, then splitting this into two, and freezing and thawing one aliquot. This should be avoided, however, since the dilution owing to the additions will compromise activity, probably by destabilizing a proportion of the membranes present.

7. RNase A purity varies considerably according to source and batch, so the conditions quoted here should be regarded as a starting point for titration. To titrate a batch of enzyme, prepare an extract, and treat aliquots of extract with a range of final ribonuclease concentrations from 0.1–5 µg/mL. Neutralize these with ribonuclease inhibitor as described, and then perform translation reactions on each in the presence and absence of a well-characterized synthetic mRNA. SDS-PAGE analysis should show a declining background with rising RNase concentration, whereas, at high concentrations, the signal resulting from the

added mRNA will be lost. Choose the "crossover" point where the signal-to-noise ratio is highest for future use. This concentration should be appropriate for subsequent extracts. Ensure that as much as possible of the batch of ribonuclease is then stored in aliquots, so that this procedure does not have to be repeated too often.

8. Dilution of the reaction mixture beyond a final volume of about 130% of the volume of extract present causes a reduction in activity, so in general, any additions to the extract, including mRNAs, should be made in the smallest volume possible. If a significant degree of dilution is a necessary part of the experiment, then wherever possible, reagents to be added should be dissolved in 10% sucrose to ensure membrane stability.

9. Generally, the performance of mRNAs from different sources parallels that seen in the *Xenopus* oocyte. Most mRNAs derived from higher eukaryotes translate well, whereas those from prokaryotes are not as effective. As with any other translation system, natural mRNAs are more efficiently translated. We find that, almost invariably, synthetic mRNAs are translated more efficiently if they are transcribed from the vector pSP64T *(3)*. A reliable protocol for transcription of synthetic mRNA is described in **ref.** *1*.

10. The amount of mRNA to be added to a translation reaction depends largely on its origin and activity. In general, a final concentration of around 50 µg/mL of synthetic mRNA produced from a cDNA cloned into pSP64T or 100–200 µg/mL of poly A$^+$ mRNA gives maximal signal without saturating the capacity of the extract to modify the translation products posttranslationally. Beyond these levels, the efficiency of segregation of secretory proteins begins to decline, and N-glycosylation becomes less efficient. Translocation without signal sequence cleavage has, however, not been observed, even at very high mRNA levels.

11. The methionine pool of the extract is approx 35 µM (± 10%). The quantity of protein produced in a reaction can therefore be estimated from the percentage incorporation of ^{35}S-methionine into TCA-precipitable material. When a reaction is programmed with a single cloned mRNA, the methionine content of the product will be known. In the case of a complex mixture, such as poly A$^+$ mRNA, being used, a reasonable estimate of average methionine content is 2%. Incorporation of all the methionine present in the extract would therefore indicate that 240 µg/mL of protein had been synthesized.

   If the aim of a reaction is to produce the largest possible quantity of protein, the yield can be increased by around 50% by the addition of excess amino acids. This is achieved by adding 5% by volume of a solution containing 700 µM methionine and 2 mM of all other amino acids to the extract, since increasing the methionine concentration more than twofold has no further effect and only serves to reduce further the specific activity of the radiolabel.

12. ^{35}S-methionine is used routinely to follow translation reactions, because it is readily available and generates good signals owing to the relatively low level of methionine in the extract. Some proteins, however, contain little or no methionine, and so require an alternative strategy. Since the pool size for methionine in the extract is comparable with that measured for the *Xenopus* oocyte, it is probably

reasonable to expect the same to apply to other amino acids. Amino acids with small pool sizes in the oocyte include cysteine, leucine, histidine, and proline, whereas there are very large pools of lysine, aspartic acid, threonine, serine, glutamic acid, glycine, and alanine. *See* **ref.** *4* for a more detailed list.

The specific activity of any amino acid can be increased by simply adding more radioactive material, but the effect on the final dilution of the extract in the reaction should be considered and, if necessary, the radiolabel should be concentrated before use. Tritiated amino acids are often supplied as dilute solutions and invariably require concentration. Generally, it is not advisable to use crude extracts of mixed radioactive amino acids, such as Translabel, in cell-free translation systems.

13. If fractionation or protease protection experiments are planned, these should, ideally, be performed immediately after the translation reaction, but we have produced acceptable results from reactions that have been stored at –70°C after freezing in liquid nitrogen.

14. We normally use the Bio-Rad Mini Protean II apparatus to analyze translation products by SDS-PAGE, using the reagents recommended by the manufacturer. The high protein content of the extract limits the amount of material that can be loaded to the equivalent of 1 µL of whole extract/5-mm wide slot on a 0.75-mm thick gel. Sucrose gradient pellets contain about a fifth of the protein present in the whole extract, and the proportion loaded can be increased accordingly, if this is not restricted by the need to load comparable amounts of unfractionated material. We routinely treat gels containing ^{35}S-labeled proteins with En3Hance (DuPont) before drying and exposure to Kodak XAR 5 film at –70°C.

15. In protease protection experiments, some transmembrane proteins with cytoplasmic tails of significant length will show a reduction in size owing to trimming by the protease.

References

1. Matthews, G. M. and Colman, A. (1991) A highly efficient, cell-free translation/translocation system prepared from *Xenopus* eggs. *Nucleic Acids Res.* **19,** 6405–6412.

2. Murray, A. W. (1991) Cell cycle extracts, in *Methods in Cell Biology*, vol. 36 (Kay, B. K. and Peng, H. B., eds.), Academic, San Diego, pp. 581–605.

3. Kreig, P. A. and Melton, D. A. (1984) Functional messenger RNAs are produced by SP6 in vitro transcription of cloned cDNAs. *Nucleic Acids Res.* **12,** 7057–7070.

4. Colman, A. (1984) Translation of eukaryotic messenger RNA in *Xenopus* oocytes, in *Transcription and Translation—A Practical Approach* (Hames, B. D. and Higgins, S. J., eds.), IRL, Oxford, pp. 271–302.

33

Purification and Characterization of Viral dsRNA Genome Profiles by Crosshybridization

Lesley-Ann Martin and Peter P. C. Mertens

1. Introduction

Double-stranded (ds) RNA present as genetic elements are widely distributed in nature and include genomes of viruses that can infect both eukaryotic and prokaryotic cells. These include members of the family *Reoviridae, Birnaviridae, Partitiviridae, Hypoviridae, Totiviridae,* and *Cytoviridae,* yeast killer factors and virus-like particles occurring as dsRNA plasmids in fungi *(1).* In most cases, the genomes of these viruses are segmented and can be separated into different bands when analyzed by SDS-PAGE or agarose gel electrophoresis, making them ideal candidates for analysis by crosshybridization. The advantage of this method is that it allows a rapid comparison of the sequence homology in every segment of the whole viral genome, of several viruses at once. Although this would also be possible, and a more definitive result can be obtained by full sequence analysis, it inevitably takes much longer particularly in the case of viruses that contain 10–12 genome segments (*Reoviridae*). The data obtained using this method allows individual homologous genome segments to be specifically identified and compared (using probes made from either isolated dsRNA segments or separate cDNA clones) (**Fig. 1**). Similarly the levels of sequence variation in different genome segments of different virus isolates can be correlated with variations in serotype and geographical origin *(2–6).*

This chapter outlines three methods that form the basis of this protocol, two of which (purification and pCp labeling) can be used independently for cloning and diagnostic purposes in which limited amounts of

From: *Methods in Molecular Biology, Vol. 86: RNA Isolation and Characterization Protocols*
Edited by: R. Rapley and D. L. Manning © Humana Press Inc., Totowa, NJ

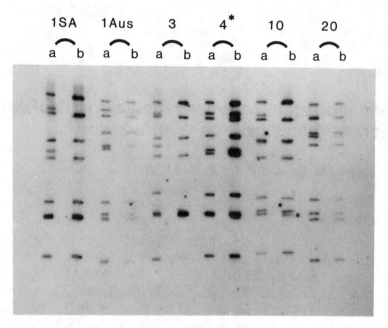

Fig. 1. Samples of BTV genomic dsRNA were separated by electrophoresis using a 10% polyacrylamide gel with a Laemmli buffer system. Each lane a was loaded with approx 3000 cpm of the appropriate BTV genomic dsRNA (as indicated), which had been labeled with [^{32}P]pCp and RNA ligase. Each lane b was loaded with 50 ng of unlabeled genomic RNA of the same BTV isolate (as indicated). After electrophoresis the dsRNA was denatured in the gel, transferred to DPT paper, and probed in this case with a probe made from BTV4 genomic dsRNA. The positions of genome segments 2 and 5 (BTV1 (Aus), 2 and 6 in lanes b are indicated by arrows. These segments have been shown to encode the outer capsid proteins which control virus serotype *(8)*. (Used with permission from **ref. *3*.)

dsRNA are available. Similarly, they are useful for studying viral genome reassortment, a characteristic trait exhibited by the *Reoviridae (7,8)*.

1.1. RNA Purification

The purification technique used to extract genomic dsRNA from orbiviruses (adapted to cell culture) involves phenol:chloroform extraction (to remove viral and cellular proteins) followed by differential precipitation of viral dsRNA and cellular ssRNA in the presence of lithium chloride. The yield of dsRNA extracted varies between 20–50 µg per 2×10^7 cells and is suitable for the preparation of cDNA by RT-PCR (which can be used in cloning strategies) and for diagnostic and comparative studies.

1.2. Labeling

1.2.1. [5'-³²P] pCp Labeling

The radiolabeling of the genomic dsRNA segments to provide electrophoretic migration markers for this protocol, relies on the addition of radiolabeled phosphate to the 3' OH termini of the dsRNA, by using T4 RNA ligase and pNp (3',5'-cytidine [5'-³²P] diphosphate pCp). Although pCp radiolabeling can be used to label small amounts of dsRNA (for diagnosis and characterization of different ds RNA viruses electropherotypes *[4]*), it is not ideal for the preparation of radiolabeled probes for crosshybridization studies. The ligase adds only a single radionucleotide to each strand of RNA. Therefore, as a function of their length, the smaller genome segments will contain relatively more label per base than larger segments. Any attempt to evaluate overall homology will inevitably be skewed towards the values for the smaller genome segments. The 3' end radiolabeling may also show some bias towards conservation levels at or near the 3' termini, particularly if there is a significant level of probe degradation during denaturation, hybridization, or washing procedures. In view of the ubiquitous nature of RNase, and because of the conserved group of bases found at the termini of many of the dsRNA viruses *(9),* together with higher levels of conservation that are found in the near terminal noncoding regions, this may pose a problem.

1.2.2. Alkaline Fragmentation and Polynucleotide Kinase Labeling

To avoid the anomalies outlined in **Subheading 1.2.1.**, the probes for crosshybridization studies can be prepared by partial degradation of purified genomic dsRNA with alkali, followed by labeling the 5' OH ends with ³²P using polynucleotide kinase. This incorporates radioisotope at many sites along both strands of the molecule, and radiolabeling will therefore be approximately proportional to the fragment length. The signal obtained in blotting experiments is therefore a function of the amount of RNA sequence present (i.e., the size of each genome segment) and the level of homology. The size of the probe molecule can also be controlled by using size exclusion columns allowing some estimate of the correct crosshybridization conditions to be calculated for given levels of stringency/homology detection.

With the increasing availability of cDNA clones and sequence data of genome segments of some dsRNA viruses, DNA probes of individual segments can be prepared using standard DNA radiolabeling techniques *(2,5,10)*. However, for uncharacterized viruses, the initial cloning steps to reach this point are lengthy and time-consuming and as a consequence the following protocol is more advantageous.

1.3. RNA Transfer and Crosshybridization

The Northern blotting technique used to transfer the dsRNA to nylon membrane is not based on capillary transfer (as seen with traditional Southern and Northern blotting) but utilizes transblotting equipment normally associated with protein transfer. In this protocol, the dsRNA is extracted and divided into two aliquots, one radiolabeled with pCp (marker lanes), the other unlabeled. These are loaded alternately onto a polyacrylamide gel so that each pair of lanes represents one serotype/isolate. Once run, the dsRNA profiles can be transferred to DPT paper or nylon membrane, hybridized with the probe and compared with its corresponding pCp marker lane. The genome segments that are homologous with the probe can be determined. By varying the stringency, the degree of sequence homology between the probe and bound target can be assessed.

2. Materials

All solutions should be prepared using molecular-grade chemicals and filter-sterilized or autoclaved where applicable. There is no need to treat solutions with DEPC. Chemicals used in thei protocol are purchased from Sigma, Poole, Dorset, U.K. unless otherwise stated.

2.1. Isolation and Rapid Purification of Viral dsRNA from Tissue Culture

1. NP40 Lysis Buffer: 120 mM NaCl, 50 mM Tris-HCl (pH 8.0), and 0.5% Nonidet P40.
2. 1 mM EDTA, pH 8.0.
3. 8M LiCl and 4M LiCl.
4. 3M Sodium acetate pH 5.2.
5. TE: 10 mM Tris-HCl, and 1 mM EDTA, pH 8.0.

All the above solutions should be autoclaved on liquid cycle and stored at room temperature.

6. Phenol:chloroform:isoamyl alcohol (25:24:1) stored at 4°C.
7. 100% and 70% ethanol.

2.2. RNA Ligase 3' Labeling of Genomic dsRNA Electrophoretic Markers

1. 10X RNA Ligase buffer: 0.5M Hepes pH 8.3, 50 μM ATP, 120 mM MgCl$_2$, 30 mM DTT. Store in aliquots at –20°C.
2. Glycerol (sterilized by autoclaving) stored at room temperature.
3. Deionized DMSO stored over amberlite mixed bed resin (sterilized by autoclaving) stored at room temperature.
4. 3',5'-cytidine [5'-^{32}P] diphosphate pCp (0.185M bq) (Amersham LIttle Chalfont, Bucks, UK, UK) (see **Note 1**).
5. T4 RNA ligase (Boehringer-Manheim, Lewes, UK or Gibco-BRL).

2.3. Polynucleotide Kinase 5' Labeling of dsRNA Probe with γ-[³²P] ATP

1. 1M NaOH.
2. 1M HCl.
3. 1M Tris-HCl, pH 8.0.
4. 10X Polynucleotide kinase buffer : 0.5M Tris-HCl (pH 7.6), 0.1M MgCl$_2$, 50 mM DTT, 1 mM spermidine, 1 mM EDTA, pH 8.0. Store in aliquots at –20°C.
5. Polynucleotide kinase (10,000 U/mL, New England Biolabs, Beverly, MA).
6. γ-[³²P] ATP (sp act 400–3000 Ci/mmole; 10 µCI/mL) 1.85M Bq approx 50 µCi (*see* **Note 1**).
7. 10X TAES: 400 mM Tris, 200 mM sodium acetate, 10% SDS, 10 mM EDTA (pH 7.8).
8. G50 spun column (nen-sorb Dupont, Stevenage, Herts, UK) (*see* **Note 2**).

2.4. Reagents for Preparation of Diazophenylthioether (DPT) Paper

1. Whatman 540 paper cut to an appropriate size (e.g., 14 × 14 cm).
2. 0.5M NaOH containing 2 mg·mL sodium borohydride (NaBH$_4$).
3. 1,4-Butanediol diglycidyl ether.
4. 2-Aminothiophenol.
5. Acetone.
6. 0.1M HCl and 1.2M HCl.
7. 1M NH$_4$OH.
8. 20% sodium dithionite in water.
9. 10 mg·mL NaNO$_2$ in water prepared immediately before use.

2.5. Electrophoresis and Northern Blotting

1. 5X SDS gel-sample buffer: 250 mM Tris-HCl (pH 6.8), 500 mM DTT, 10% SDS (electrophoresis–grade), 0.5% bromophenol blue, 50% glycerol. Store in aliquots at –20°C.
2. Tris-glycine-SDS running buffer: 25 mM Tris, 250 mM glycine (pH 8.3), 0.1% SDS.
3. 0.1M NaOH.
9. Sodium phosphate buffer (pH 5.5), 500 mM, 50 mM, and 25 mM.
5. Diazophenylthioether (DPT) paper (*see* **Note 3**).
6. Whatman 3MM paper and Autoradiographic film: e.g., Kodak XARS.

2.6. Hybridization

1. 20X SSC: 175.3 g NaCl, 88.2 g sodium citrate dissolved in 900 mL of distilled H$_2$O, adjust pH to 7.0 with NaOH make up to 1 L and sterilize by autoclaving. Store at room temperature.
2. 50X Denhardt's reagent: 5 g Ficoll (Type 400, Pharmacia), 5 g polyvinylpyrrolidone, 5 g bovine serum albumin (Fraction V; Sigma), and distilled H$_2$O to 500 mL. Filter sterilize and store in aliquots at –20°C.

3. Prehybridization buffer: 5X SSC, 50 m*M* sodium phosphate (pH 6.5) 0.1% SDS, 5X Denhardt's reagent, 100 µg/mL denatured, fragmented salmon sperm DNA (Sigma type III sodium salt) (*see* **Note 4**) and 1% (w/v) glycine (*see* **Note 5**). Store at 4°C.

3. Methods

3.1. Extraction of dsRNA from Small-Scale Tissue Culture

1. Harvest a 175-cm^2 flask of BHK cells (or an appropriate cell line used for virus propagation) when full cytopathic effect (CPE) is observed (a cell scraper can be used to remove any cells which adhere to the plastic). Pour the cell suspension into two universals. Spin at 250g$_{ave}$ for 10 min at 4°C. Pour off the supernatants.
2. Pool and resuspend pellets in 400 µL of NP40 buffer (pH between 5.0 and 6.0) incubate at room temperature for a minimum of 30 min mixing occasionally by swirling.
3. Add 0.5 mL of phenol vortex and incubate at 60°C for 5 min. Then place on ice for a further 5 min.
4. Spin the sample for 2 min at 13,000g (microfuge) remove the aqueous phase and add 0.5 mL of phenol:chloroform:isoamyl alcohol (25:24:1). Vortex for 2 min then spin 2 min.
5. Remove the aqueous phase and add 0.5 mL of ether. Spin for 2 min.
6. Discard the ether (upper phase) and add 1 mL of 100% ethanol and 20 µL of 3*M* sodium acetate. Precipitate at –70°C for 1 h or overnight at –20°C.
7. Spin for 15 min, discard the ethanol and resuspend the pellet in 300 µL 1 m*M* EDTA.
8. Add 100 µL of 8*M* LiCl and leave at 4°C for a minimum of 8 h.
9. Spin at 13,000g for 10 min, discard the pellet and add 200 µL of 8*M* LiCl to the supernatant, leave for a minimum of 8 h at 4°C.
10. Spin the sample for 10 min at 13,000g and discard the supernatant. The dsRNA can be seen as small granules about the size of grains of sand on the side of the Eppendorf.
11. Add 0.5 mL of 100% ethanol and 50 mL 3*M* sodium acetate to the tube and precipitate at –70°C for 1 h then spin the dsRNA at 13,500 rpm for 15 min.
12. Wash the pellet in 70% ethanol spin down as before for 5 min and then dry the pellet.
13. Resuspend the dsRNA pellet in 50 µL of TE, run a 2 µL sample on a 1% agarose gel against 1 µg of 1 Kb ladder (Gibco-BRL) to quantify and check the integrity of the viral dsRNA. A 175-cm^2 flat will yield between 20–50 µg of bluetongue or African horse sickness viral dsRNA, this can be verified by OD$_{260}$ readings.
14. The dsRNA can be stored in 70% ethanol at –20°C for several years without signs of degradation.

3.2. Preparation and Testing of Diazophenylthioether (DPT) Paper

Carry out all operations in fume hood and wear gloves as many of the reagents are extremely toxic.

1. Take 7 sheets of Whatman 540 paper (14 × 14 cm) and place in a heat sealable bag or glass roller bottle.
2. Add 70 mL of $0.5M$ NaOH containing 2 mg·mL sodium borohydride ($NaBH_4$).
3. Pour in 30 mL of 1,4-butanediol diglycidyl ether and seal the bag, leaving enough room to cut and reseal the bag later.
4. Agitate by rotating the bag end-over-end for 8–16 h at room temperature.
5. Open the bag and pour off the liquid into a beaker. Before disposing the 1,4-butanediol diglycidyl ether, add it to $1M$ NH_4OH and leave for 24 h before pouring down a sink.
6. Add 10 mL of 2-aminothiophenol in 40 mL of acetone. Reseal the bag and agitate for a further 10 h as described in **step 4**.
7. Open bag and remove the paper.
8. Wash twice in acetone, twice in $0.1M$ HCl, and copiously with distilled H_2O. Wash once more with $0.1M$ HCl, then distilled H_2O and air dry.

To check that the DPT paper is activated the following steps should be carried out.

9. In a fume hood incubate the DPT paper in 150 mL of 20% sodium thionite in H_2O at 60°C for 30 min with occasional swirling.
10. Wash the DPT paper with a large volume of water (1–2 L).
11. Wash once with 100–200 mL of $1.2M$ HCl for 14 × 14 cm paper.
12. Transfer paper directly to 0.3 mL/cm² of ice cold $1.2M$ HCl. For each 100 mL of HCl add, with mixing, 2.7 mL of a solution of $NaNO_2$ in water (10 mg·mL) prepared immediately before use.
13. Keep the paper in this solution on ice for 30 min or longer with occasional swirling.
14. After 30 min, a drop of solution should give a positive (black) reaction with starch iodide paper.
15. Leave the paper in ice-cold $1.2M$ HCl until needed.
16. Pour off the acid, wash the paper for no more than 2–3 min, twice in ice cold H_2O and twice in ice–cold transfer buffer (25 mM sodium phosphate buffer, pH 5.5). The paper should now be bright yellow (*see* **Note 6**).

3.3. Labeling of the 3′ Termini of Viral dsRNA with T4 RNA Ligase

This section outlines the labeling of viral genomic dsRNA with [5′-³²P]-pCp for use as electrophoretic migration markers *(12)* (*see* **Note 7**).

1. To 1 μg of dsRNA, add 6 μL of 10X T4 RNA ligase buffer, 10% deionized DMSO (6 μL), 15% glycerol (9 μL), $0.185M$ Bq (5 μCi) [5′-³²P]-pCp (2.5 μL), 5 U of T4 RNA ligase and make up to a final volume of 60 μL. Incubate at room temperature for 4 h or overnight at 4°C (*see* **Note 8**).
2. After incubation add 40 μL of sterile distilled H_2O to the radiolabeled ds RNA and store at –20°C.
3. Make serial dilutions of a small aliquot of the pCp-labeled markers in 2X SDS-sample buffer, heat for 3 min at 50°C and run them on a 10% SDS-PAGE

gel (*see* **Notes 9** and **10**). The gel (still attached to the glass plate) should be covered in Saran Wrap and immediately exposed to film. After an overnight exposure the dilution giving the clearest autoradiographic profile can be used in subsequent experiments (this is approximately equivalent to 3000 cpm).

3.4. Preparation of Genomic Viral dsRNA Probe

The probe consisting of total genomic viral dsRNA is prepared by alkaline fragmentation followed by labeling with [γ-^{32}P]-ATP (*see* **Note 11**).

1. Boil 20 µg of total genomic dsRNA (or 5 µg if a single genome segment is used) in 100 µL of distilled H$_2$O for 2 min, then cool on ice for 2 min.
2. Add 10 µL of 1*M* NaOH and incubate on ice for 10 min, then add 20 µL of 1*M* Tris-HCl (pH 8.0), 13 µL of 3*M* sodium acetate pH 5.2 and 3 vol of ice-cold 100% ethanol. Precipitate the RNA at –70°C for 15 min. Centrifuge the sample at 13,000*g* for 20 min. Discard the ethanol and wash the RNA pellet with 1 mL of 70% ethanol. Spin for 5 min as described above, discard the ethanol and dry the RNA.
3. Resuspend the alkali digested RNA in 20 µL of distilled H$_2$O, remove an aliquot corresponding to 1 µg (the remaining sample can be ethanol precipitated and stored at 20°C).
4. Add to the 1 µg aliquot of dsRNA 5 µL of 10X polynucleotide kinase buffer, 2 µL of polynucleotide kinase (20U), 100µCi γ-[^{32}P] ATP and make up to a final volume of 50 µL with H$_2$O. Incubate the sample at 37°C for 1 h.
5. Place the sample on ice and add 50 µL of 1X TAES. Apply the sample to a G50 Nen-Sorb spun column according to the manufacturer's instructions. Remove a 1-µL aliquot and measure the activity by means of a scintillation counter (specific activity should be approximately 1 × 10^8 cpm/µg. The radiolabeled probe can be stored at –20°C until required (*see* **Note 12**).

3.5. Gel Electrophoresis and Northern Blotting of Genomic dsRNA

1. Prepare a 10% SDS-polyacrylamide gel with 4% stacking gel (**Table 1**).
2. Load each lane with approx 3000 cpm of the appropriate pCp-labeled genomic dsRNA marker and each lane with 50 ng of corresponding unlabeled dsRNA (**Fig. 1**).
3. Run the polyacrylamide gel in Tris-glycine running buffer at a constant 216V overnight (at least 16 h). Snip the right hand-corner of the gel to orientate it.
4. Soak the gel in 0.1*M* NaOH for 15 min (this denatures the RNA and reduces the average molecular weight improving elution) (*see* **Note 13**).
5. Wash the gel in 500 m*M* sodium phosphate buffer (pH 5.5) for 10 min at 0°C. Repeat this wash step with 50 m*M* phosphate buffer (pH 5.5) and 25 m*M* phosphate buffer (pH 5.5). The gel is now ready for transfer. We use a Trans-blot electrophoretic transfer cell (Bio-Rad) according to the manufacturer's instructions.
6. Rinse the DPT paper in 25 m*M* and assemble the blot and insert into the electrophoretic transfer cell.

Table 1
Gel Recipes for 10% SDS-Polyacrylamide Running Gel
and 5% Stacking Gel *(11)*

	Running gel		Stacking gel
1. 30% bis/acrylamide	10 mL		1.7 mL
2. 1.5*M* Tris, pH 8.8	7.5 mL	1M Tris, pH 6.8	1.25 mL
3. 10% SDS	0.3 mL		0.1 mL
4. H$_2$O	11.9 mL		6.8 mL
5. 10% ammonium persulphate	0.3 mL		0.1 mL
6. TEMED	0.012 mL		0.01 mL

7. Electrophoretic transfer is carried out in 25 m*M* sodium phosphate buffer (pH 5.5) at 4°C, 10 V/cm and 2.5 A for 4 h.
8. The blot can be immediately used or blotted dry, and stored at –70°C.

3.6. Hybridization and Washing Conditions

The hybridization and stringency washing conditions can be calculated using a modification of the formula described by Bodkin and Knudson *(13)* which reduces the calculated RNA T$_m$ by (650/[average probe length in bases]) *(14)*.

$$T_m = 79.8°C + 18.5 (\log_{10} [Na+]) + 0.58 (\text{fraction G+C}) + 11.8 (\text{fraction G+C})^2 - 0.35 (\% \text{ formamide}) - 650/l$$

l = length of the probe in base pairs (probes produced by this method are estimated at 150 bp) (*see* **Note 14**).

The conditions outlined below for the stringency washing assume the percentage of G+C bases in the dsRNA viral genome to be 42.4% *(15)* (*see* **Note 15**).

1. Add 0.2 mL of prehybridization buffer containing 50% formamide per cm^2 of membrane. Incubate at 50°C for 4 h.
2. Immediately prior to hybridization, boil the probe for 5 min, then chill on ice. Add the probe to the hybridization buffer at approx 80–100 ng/mL. Remove the prehybridization buffer and replace with the probe/hybridization buffer. Hybridize for 16–24 h at 50°C.
3. Discard the probe and wash the filter four times for 5 min in 2X SSC + 0.1% SDS at room temperature.
4. Wash the filter at high stringency in 0.1X SSC + 0.1% SDS for 15 min at 58°C with two changes.
5. Lay the membrane on a square of Whatman paper and cover with Saran Wrap or cling film. Expose the membrane to Kodak XAR film and store at –70°C. Develop the film after 24–48 hours (**Fig. 1** shows typical results).

4. Notes

1. Standard precaution when using radioactive isotopes should be observed.
2. G50 spun columns that are preprepared are often expensive and can be easily made as follows. Resuspend Sephadex G50 (Pharmacia, Uppdala, Sweden) in 1X TAES to provide a slurry, leave this to rehydrate overnight at room temperature. Using gloves, make small balls of glass wool approximately the size of a pea and dip these in Replicote (Sigma) (this prevents the probe sticking to the glass). Place the plugs in a small beaker and autoclave them. Store at room temperature. Remove the plunger from a 1-mL syringe and plug the base of the barrel with a glass wool plug. Fill the syringe with the sephadex G50 slurry, allowing the TAES to drain out, keep topping up the syringe until it contains 1 mL of G50. Place the syringe into a 15-mL Falcon tube (2095) and spin for 2 min at $250g_{ave}$. The column will look dry and cracked. Remove it with a pair of forceps. Place a screw cap Eppendorf tube in the base of a new Falcon tube and insert the column into the Eppendorf tube. Load the radioactive probe onto the center of the column, seal the Falcon tube, and spin for 2 min as before. Remove the column and the Eppendorf tube using forceps. The Eppendorf tube now contains the purified labeled probe. Discard the column as radioactive waste.
3. Zeta-probe nylon membrane (Bio-Rad, Hemel, Hempstead, Herts, UK) can be used instead of DPT paper for the Northern transfer step. Some researchers may find this a more acceptable and less toxic alternative! However, we have encountered problems in the efficient transfer of some dsRNAs to nylon membranes and in these cases the DPT that covalently binds the transferred RNA was found to be superior.
4. Denatured fragmented Salmon sperm DNA can be prepared by dissolving the DNA in water to give a final concentration of 10 mg·mL (this may require the aid of a magnetic stirrer over a 2–4 h period at room temperature). The solution should then be adjusted to $0.1M$ sodium chloride and extracted twice with phenol:chloroform. The aqueous phase containing the DNA should be sheared by passing it several times through a 17-ga hypodermic needle and precipitated with 2 vol of ethanol, then centrifuged for 15 min at $13000g_{ave}$. Wash the pellet with 70% ethanol, air dried and resuspended in water to give a final concentration of 10 mg·mL (determined by its OD_{260}). Boil the solution for 10 min and store in aliquots at $-20°C$.
5. Glycine 1% (w/v) is used in the prehybridization buffer to inactivate the DPT residues after RNA transfer. If nylon membrane is used for blotting, glycine is omitted from the prehybridization solution.
6. The DPT paper is stable for several months at room temperature, but must be protected from light and alkali reagents.
7. The pCp-labeling method can be used to label as little as 100 ng of viral genomic dsRNA for diagnostic profile comparisons and identification.
8. The pCp-labeling technique relies on the addition of phosphate groups to the dsRNA 3' termini by T4 RNA ligase. This enzyme is most efficient at catalyzing the reaction with ssRNA. If the incorporation is low, this is sometimes due to inhibition of the enzyme by residual phenol. To overcome this the dsRNA stock

should be reprecipitated with 100% ethanol in the presence of sodium acetate and the resulting pellet washed twice with 70% ethanol.

9. The pCp-radiolabeled markers can be separated from unincorporated label by running through a G50 spun-column as described in **Subheading 3.4., step 5** and the activity measured by scintillation counting. However, visual analysis by SDS-PAGE as described in **Subheading 3.3.** is recommended.

10. It should be noted that if the unincorporated pCp-label is not removed from the markers as described in **Subheading 3.3.**, then the resulting electrophoresis buffers must be treated as radioactive.

11. The method can be modified so that one genome segment can be isolated by agarose or polyacrylamide gel electrophoresis purified, labeled and used as a single probe.

12. The labeled probes are generally stable for 1–2 wk but their use is dependent on their radioactive half life. Northerns hybridized with probes that have past one half-life give weak signals and autoradiographs need to be exposed for longer periods.

13. If Zeta-probe nylon membrane is to be used for the Northern transfer, proceed with the denaturing step described in **Subheading 3.5.**, then rinse the gel twice for 10 min in 5X TAE buffer, and then 10 min in 0.5X TAE. Transfer is carried out in 0.5X TAE at 0.8A for 2 h. Once the transfer is complete mark the membrane with a pencil to orientate it. Then fix the dsRNA to the membrane by exposing it to UV light for 5 min on each side. Alternatively the dsRNA may be fixed by baking at 80°C for 2 h in a vacuum oven.

14. The alkaline hydrolysis produces probes approx 150 bp in length. This arbitory size can be used to calculate the conditions for hybridization and stringency washes. If necessary, the length of the probe can be checked by running a small aliquot on a 1.5% agarose gel against a radiolabeled DNA marker. An overnight exposure of the gel should reveal a smear of fragmented RNA that can be measured against the DNA marker.

15. The protocol described was optimized for crosshybridization studies between bluetongue virus isolates. Under the high stringency conditions employed the T_m (RNA) was calculated as 70.7°C. The blot was washed at 58°C (which is a T_m (RNA) 12.7°C lower). Assuming a 1% mismatch in the RNA sequences reduces the T_m of the hybrids by 1.4°C; stable hybrids would be formed that contain approx 90% sequence homology and 9% mismatch. A variation of up to 4.5% in the G+C content of the viral genome segments would alter the sequence homology required for stable duplex formation by only 2.2%. The hybridization conditions for other viral genomes can be calculated similarly allowing degrees of homology to be determined.

References

1. Murphy, F. A., Fauquet, C. M., Bishop, D. H. L., and Ghabrial F. A., eds. (1995) *Virus taxonomy: Classification and Nomenclature of Viruses, 6th Report of the International Comittee on Taxonomy of Viruses.* pp. 205–264.

2. Street, J. E., Croxson, M. C., Chadderton, W. F., and Bellamy, A. R. (1982) Sequence diversity of human rotovirus strains investigated by northern blot hybridization analysis. *J. Virol.* **43(2)**, 369–378.

3. Mertens. P. P. C., Pedley, S., Cowley, J. ,and Burroughs, J. N. (1987). A comparison of six different virus isolates by cross-hybridization of the dsRNA genome segments. *Virology* **161**, 483–447.

4. Mertens, P. P. C., Crook, N. E., Rubinstein, R., Pedley, S., and Payne, C. C. (1989) Cytoplasmic polyhedrosis virus classification by electropherotype: validation by serological analysis and agarose electrophoresis. *J.Gen.Virol.* **70**, 173–185.

5. Bremer, W., Huismans, H., and Van Dijk, A. A. (1990) Characterization and cloning of the African horsesickness virus genome. *J. Gen. Virol.* **71**, 793–799.

6. Gould, A. R. and Pritchard, L. I. Phylogenetic analyses of complete nucleotide sequence of the capsid protein (VP3) of Australian epizootic hemorrhagic disease of dear virus (serotype-2) and cognate genes from other orbiviruses. *Virus Res.* **21(1)**, 1–18.

7. Schroeder, B. A., Street, J. E, Kalmakoff, J., and Bellamy, A. R. (1982) Sequence relationships between the genome segments of human and animal rotavirus strains. *J. Virol.* **43(2)**, 379–385.

8. Mertens, P. P. C., Pedley, S., Cowley, J., Burroughs, J. N., Corteyn, A. H., Jeggo, M. H., Jennings, D. M., and Gorman, B. M. (1989) Analysis of the roles of bluetongue virus outer capsid proteins VP2 and VP5 in determination of virus serotype. *Virology***170**, 561–565.

9. Mertens, P. P. C. and Sangar, D. C. (1985) Analysis of the terminal sequence of the genome segments of four orbiviruses.*Virology* **140**, 55–67.

10. Huismans, H. and Cloete, M. (1987) A comparison of different cloned bluetongue virus genome segments as probes for the detection of virus-specified RNA. *Virology* **158**, 373–380.

11. Laemmli, U. K. (1970) Cleavage of structural proteins during the assembly of the head of bacteriophage T4. *Nature* **227**, 680–685.

12. England, T. E and Uhlenbeck, O. C. (1978) 3' terminal labeling of RNA with T4 RNA ligase. *Nature* **275**, 560–561.

13. Bodkin, D. K. and Knudson, D. L. (1985) Sequence relatedness of Palyam virus genes to cognates or the Palyam serogroup viruses by RNA-RNA blot hybridization. *Virology* **143**, 55–62.

14. Howley, P. M, Israel, M. A., Law, M., and Martin, M. A. (1979) A rapid method for detecting and mapping homology between heterologous DNAs. *J. Biol. Chem.* **254**, 4876–4883.

15. Verwoerd, D. W. (1969) Purification and characterisation of blue-tongue virus. *Virology* **38**, 203–212.

Index